THE MPH COOKBOOK

by ZARINAH ANWAR

Photography by
Chua Soo Bin

Compiled by
Rosalind Mowe

Edited by
Wendy Hutton

Designed by
Elaine Ng

First published in Singapore by
MPH Distributors (S) Pte Ltd.

First Canadian Edition published 1979 by
Souvenir Press Ltd, 43 Great Russell
Street London WC1B3PA in association with
John Wiley & Sons Canada Ltd., 22
Worcester Road, Rexdale, Ontario, M9W ILI

ISBN 0-471-99828-1

Printed and bound in Singapore
by Toppan Printing Company (S) Pte. Ltd.

FOREWORD

ZARINAH ANWAR is in her early thirties, married and with a young son of school age. Her keen interest in cookery started when she was very young, and at the age of twelve she would watch and help her mother cook. By the time she was fourteen, her coconut candy was well known among her fellow students and teachers, as well as in her own neighbourhood.

The recipes offered to you in this book have been in the family for many years, handed down by her grandmother, mother and aunts. She learnt the skills and intricacies of North and South Indian cookery from her grandmother and mother, and an aunt taught her the secrets of Indonesian and Malay cookery. Her knowledge of Chinese cookery was gained from a Chinese aunt. She learnt to make Nonya **kueh** through close observation and experiment until she perfected each recipe. A firm believer in the old cooking methods, Mrs. Anwar is not in favour of commercial ready-made curry pastes or food additives. Given a choice, she would rather use a **batu giling** than a liquidiser.

In her modest kitchen the author prepares spices to make up her own brands of curry powder for meat, fish, kurmah and other dishes, catering for lunches and dinners among friends. Those who have tasted her food have always gone back for more.

— Rosalind Mowe

Preface

Singapore, by virtue of its multi-racial make-up, has brought about a fusion of cultures and culinary tastes. Tolerance for the food of various races has made the island a paradise for all who enjoy a good meal.The Singaporean will travel for miles if he knows that at the end of his journey he will be able to satisfy his desire for a certain delicacy or type of food; however he also delights in a good, wholesome meal cooked at home.

This book has been written for all those who savour the pleasure of unusual, delicate and enterprising home prepared meals, and who are willing to venture into the delightful and satisfying world of cookery. The dishes are drawn from many parts of Asia, but demand no magic formula, just plain down to earth common sense. They can all be prepared in your own home, within the scope of an ordinary family budget.

The book is intended for the expert cook who wants new ideas, the less-expert who needs both ideas and guidance, and the novice cook, needing step-by-step assistance. To the new cook I would say, don't be discouraged if something goes wrong along the way. Perhaps you have overlooked some point in the recipe or misread it in your anxiety. Try again, a second, a third and even a fourth time. With regular practice, you will succeed. The cook must also learn to adjust the amount of certain ingredients according to taste. Cooking is a creative art.

I wish to express my sincere gratitude to Mrs. Rosalind Mowe who worked with me for over a year testing every recipe in order to compile this book.

I sincerely hope that all those in possession of this cook book will not only have many hours of creative pleasure, but also give pleasure to all those who taste their cooking.

— Zarinah Anwar

CONTENTS

Weights & Measures

1 metric teaspoon holds 5ml
1 metric tablespoon holds 20ml
1 metric cup holds 250ml
4 metric cups=1 litre

Abbreviations

kilogram	kg
gram	g
centimetre	cm
teaspoon	tsp
tablespoon	tbsp
millilitre	ml
litre	l

WEIGHT

Local	metric equivalent	rounded metric equivalent
1 kati (16 tah)	605g	600g
¾ kati (12 tah)	454g	450g
½ kati (8 tah)	302g	300g
¼ kati (4 tah)	151g	150g
(2 tah)	76g	80g
(1 tah)	38g	40g
Imperial		
1 lb (16 oz)	453.6g	450g
¾ lb (12 oz)	340.2g	340g
½ lb (8 oz)	226.8g	230g
¼ lb (4 oz)	113.4g	120g
(2 oz)	56.7g	60g
(1 oz)	28.3g	30g

Curry Powders

FOR MEAT

Ingredients:

600 g (1 kt) coriander seeds (*ketumbar*)
150 g (¼ kt) cumin seeds *(jintan puteh)*
150 g (¼ kt) fennel seeds *(jintan manis)*
150 g (¼ kt) dried chillies
80 g (2 tah) dried turmeric root (*kunyit*)
80 g (2 tah) black peppercorns
40 g (1 tah) cinnamon sticks (*kayu manis*)
20 cloves
15 cardamoms

Method:

Remove stones and other foreign matter from spice seeds. Wash coriander, cumin and fennel and dry in the sun, then roast in a dry pan until fragrant.

Send everything to the mill for grinding and mixing (or use coffee grinder at home). Cool well after grinding before storing in airtight containers. Once a week, open the containers to air the curry powder. The length of time the curry powder lasts depends very much on the quality of the spices used.

FOR FISH

Ingredients:

600 g (1 kt) coriander seeds (*ketumbar*)
150 g (¼ kt) fennel seeds *(jintan manis)*
80 g (2 tah) cumin seeds *(jintan puteh)*
150 g (¼ kt) dried chillies
115 g (3 tah) dried turmeric root (*kunyit*)
80 g (2 tah) black peppercorns

Cooking Tips

CHICKEN:

1. If frozen chicken is used, the amount of liquid given in a recipe should be reduced a little.
2. If chicken is fat, reduce the amount of cooking oil or ghee.
3. To cook an authentic Indian chicken curry, the skin of the chicken should be removed and discarded before cooking. Chicken curry will then be less oily and will keep longer.

CHILLIES:

1. When buying dried chillies, select the crinkled variety.
2. When a recipe calls for dried chillies to be ground, always soak the chillies in water until they are soft before grinding. If hot water is used, the soaking time will be shorter.
3. For a milder flavour in a curry, remove seeds of fresh or dried chillies.
4. Add or reduce the number of chillies or the amount of chilli powder given in a recipe, according to taste.
5. If a recipe calls for dried chillies to be roasted, do this in a dry pan over *low* heat stirring now and then to prevent burning.

COCONUT:

1. When selecting coconuts, choose older ones which will yield more milk.
2. When extracting coconut milk, add *one* cup of water to grated coconut at a time and squeeze to extract milk before adding the next cup, even if recipe calls for 4 cups of water.

FRESH TURMERIC *(kunyit)*:
The older root, which is dark in colour, is preferable.

LEMON GRASS *(serai)*:
Only about 6-7 cm (2½ – 3in) of the root end should be used.

LENGKUAS *(botanical: galangal)*:
If *lengkuas* needs to be ground, buy a tender piece that is pinkish in colour.

LENTIL *(dhal)*:
Always select the larger variety; soak for about 2 hours in cold water before using, then wash and remove loose skin.

TAMARIND:

1. Select the darker variety.
2. To make tamarind juice, combine specified amount of water and tamarind pulp, mix with fingers and strain.

GENERAL:

1. If while cooking curry is drying up fast and you wish to add more water, make sure *warm* water is used.
2. Indians use yoghurt to tenderise meat; add about 1 tbsp to 450 g of meat and marinate for about 20-30 minutes before cooking.
3. One of the secrets of a good curry lies in the quality and amount of spices used. It is preferable, therefore, that you prepare your own ground spices or buy them from a reliable source. Very often, commercial curry powders are adulterated with rice or maize flour and do not contain the pure ingredients.
4. Care of the griddle *(tawa)* used for cooking chapati and Roti Jala:—
   i) Wash new griddle with warm water.
   ii) Season the new griddle by using about a handful of grated coconut. Fry this on griddle over low heat, making use of every part of the griddle. Keep stirring coconut for about 40 minutes then discard coconut. Next, heat a tablespoon of cooking oil on griddle, break an egg and spread it all over. Discard egg when it is cooked and wipe surface of griddle. It is now ready for use. After this, the griddle should *never* be washed. After use, merely wipe over with a dry cloth or a paper towel. If at any time food sticks to it during use, repeat egg treatment.

Note on Malay Spelling: the revised form of spelling introduced in 1973 has been used in this book. Thus, *blachan* is now spelled *blacan, kachang* as *kacang,* etc., but of course the pronunciation of these words remains the same.

Spice Charts

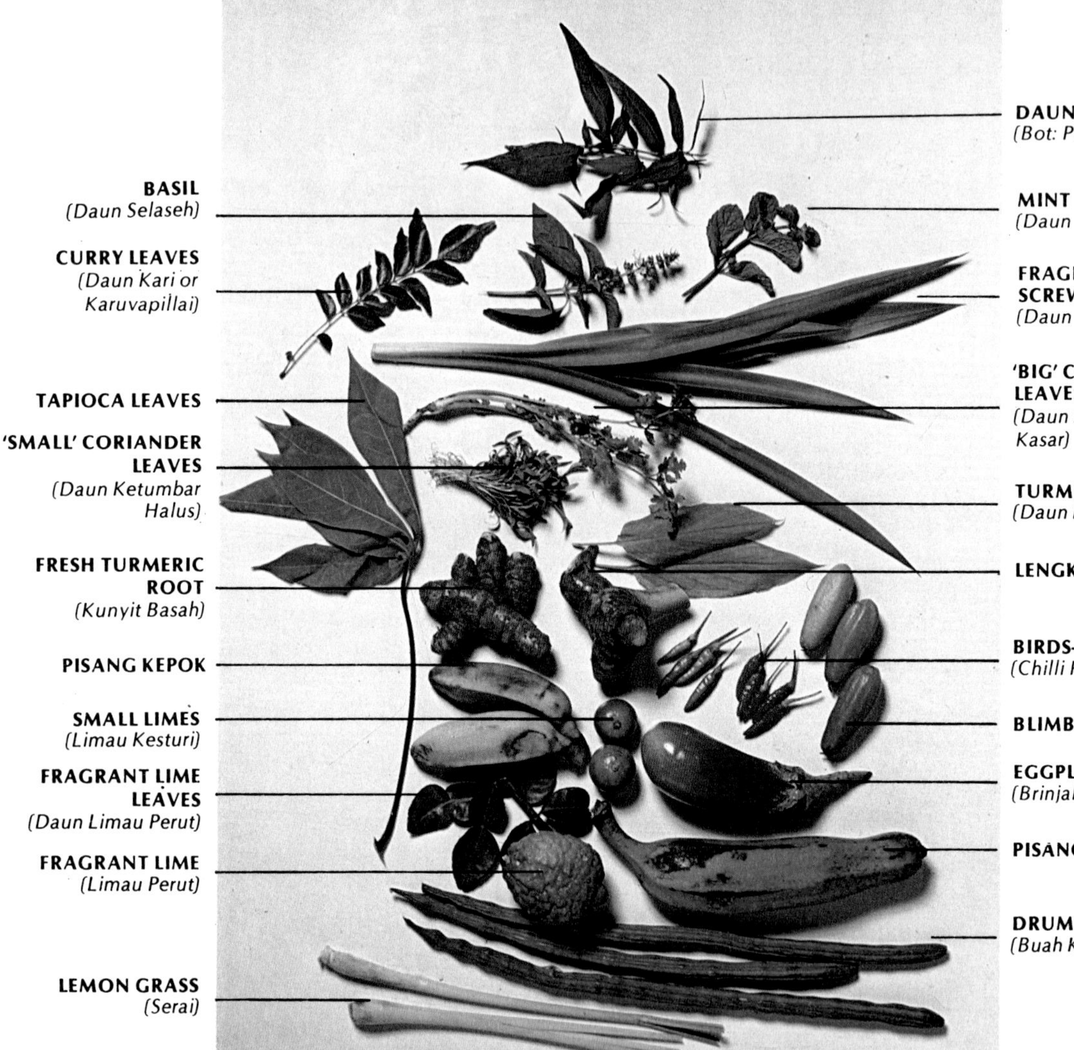

LENTILS (Dhal)
DRIED TAMARIND SLICES (Asam Keping or Asam Gelugor)
BLACK FUNGUS
DRIED BLACK CHINESE MUSHROOMS
SHALLOTS (Bawang Merah)
BEAN THREAD VERMICELLI (Sohoon)
CASHEW NUTS
WHEAT GRAINS (Terigu)
DRIED CHILLIES
CHILLIES PRESERVED IN SALTED YOGHURT (Chilli Tairu)
SMALL DRIED WHITEBAIT (Ikan Bilis)
PICKLED LIME
ALMONDS
NUTMEG (Buah Pala)
COCONUT CRUMBS (tahi minyak)
CANDLENUTS (Buah Keras)

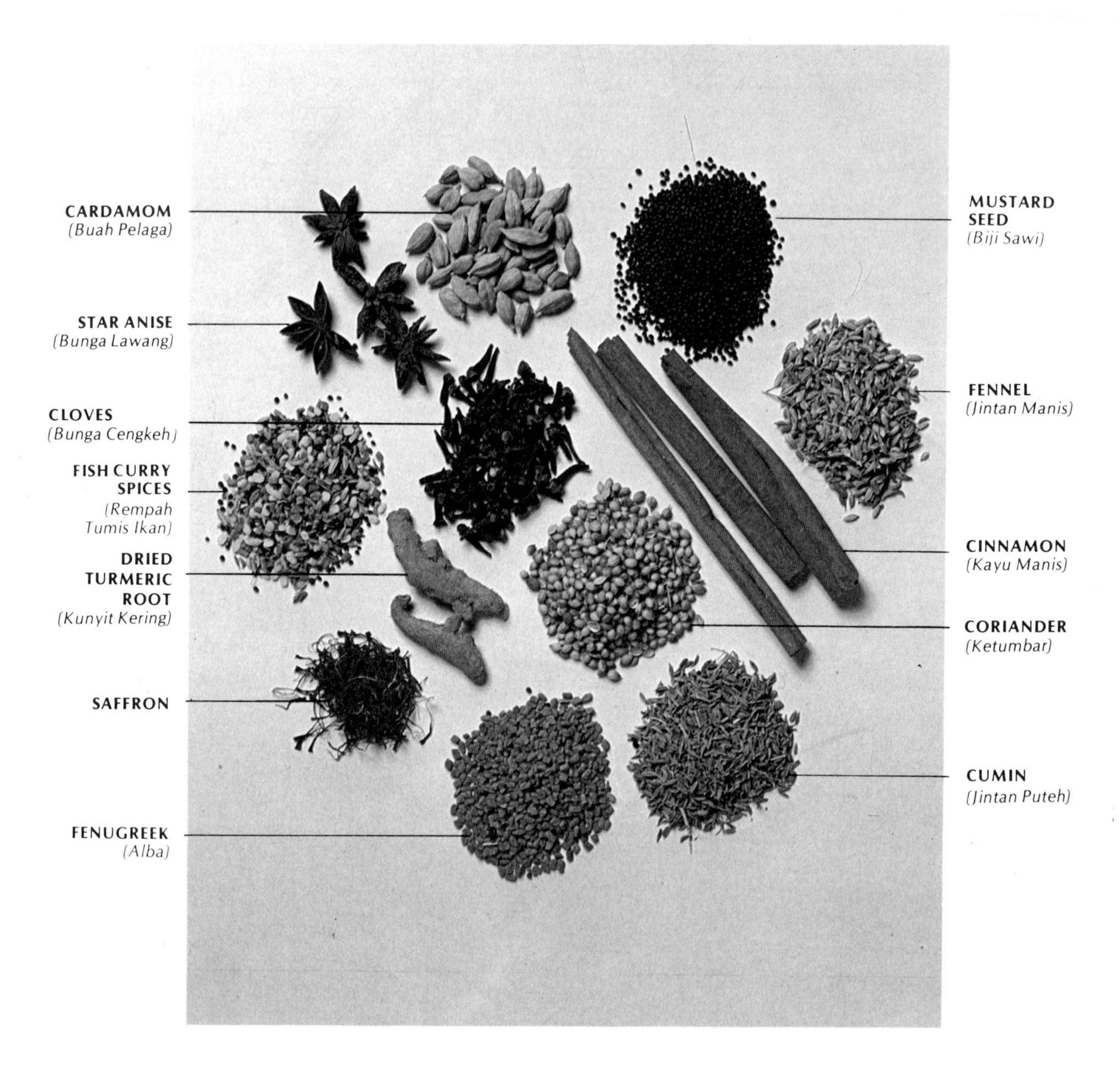
CARDAMOM
(Buah Pelaga)
STAR ANISE
(Bunga Lawang)
CLOVES
(Bunga Cengkeh)
FISH CURRY SPICES
(Rempah Tumis Ikan)
DRIED TURMERIC ROOT
(Kunyit Kering)
SAFFRON
FENUGREEK
(Alba)
MUSTARD SEED
(Biji Sawi)
FENNEL
(Jintan Manis)
CINNAMON
(Kayu Manis)
CORIANDER
(Ketumbar)
CUMIN
(Jintan Puteh)

BAMBOO STEAMER
[for Putu Bamboo]
MURUKKU MOULD
MEASURING CUP
GRIDDLE or HOT PLATE (Tawa)
BASE FOR BAMBOO STEAMER
CUP FOR ROTI JALA

ROSE WATER
(Ayer Mawar)
TOMATO PUREE
DIGA
RICE DOUGH
[Tepong Beras Boh]
GHEE
(Minyak Sapi)
PURE GHEE
MAHAL BRAND
MAWAR
CHICK PEA FLOUR
(Besan)
PAPADOM
EXTRA SUPERIOR QUALITY
ASAFOETIDA
LALJEE GODHOO & CO.
DELICIOUS TASTE
SOLE AGENT
STORES
ASAFOETIDA
(Hing or Perunkayam)

Indonesian Dishes

NASI AMBANG AND ITS ACCOMPANIMENTS

Nasi Ambang

Ingredients:
600 g (1 kt) rice – soak in water for 4–5 hours, then drain
2 fragrant screwpine leaves (*daun pandan*) – wash and tie into a bundle
2½ cups water

Method:
Boil water in steamer.
Put rice, *pandan* leaves and 2½ cups water in a deep tray and steam till rice is cooked.

Ikan Goreng

Ingredients:
4 small pieces *ikan tenggiri* (Spanish mackerel or bonito) – wash and dry
1 tsp turmeric powder
salt to taste
2 tsp water
oil for shallow frying

Method:
Mix turmeric powder, salt and water into a thick paste. Rub the pieces of fish evenly with paste and fry in oil till golden brown.

1. *Nasi Ambang* **2.** *Ayam Panggang Gahru* **3.** *Sambal Goreng* **4.** *Ikan Goreng* **5.** *Rendang Rempah* **6.** *Serundeng* **7.** *Sambal Terong* **8.** *Tempe* **9.** *Potato Cutlets* **10.** *Urap Taugeh*

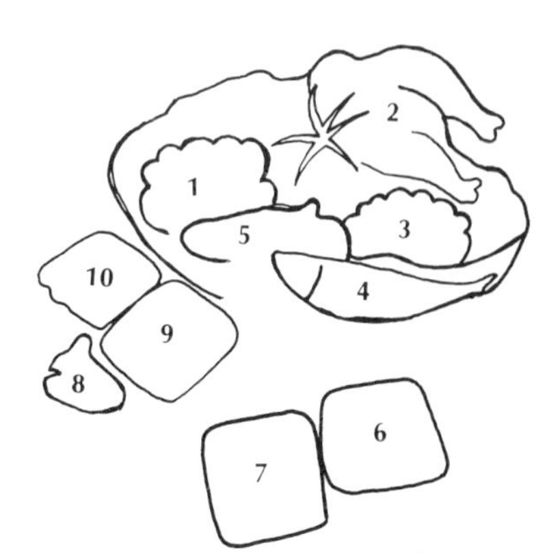

Rendang Rempah

Ingredients:

600 g (1 kt) beef or mutton – cut into small pieces
4 level tbsp coriander (*ketumbar*) powder
½ tbsp cumin *(jintan puteh)* powder
1 stalk lemon grass *(serai)* – bruise
1 piece *lengkuas* the size of a walnut – bruise
2 fragrant lime leaves (*daun limau perut*)
1 small turmeric leaf – use whole
450 g (¾ kt) grated coconut – set aside 2 tbsp to be roasted over low heat till brown. Add 5 cups water and extract 6 cups thick milk from the rest of the coconut
2 tsp sugar
salt to taste

Grind into a smooth paste:

10 dried chillies – remove seeds for a mild flavour and soak in hot water to soften
8 shallots
2 tbsp roasted coconut taken from allowance

Method:

Combine ground ingredients with meat in a deep frying pan. Add coconut milk, turmeric leaf, lime leaves, *serai, lengkuas,* coriander powder, cumin powder, sugar and salt. Bring to the boil and continue to cook uncovered till meat is tender, then reduce heat and cook till quite dry, turning over the contents from time to time to prevent burning.

Sambal Goreng

Ingredients:

300 g (½ kt) ox liver – cut into small squares
300 g (½ kt) shrimps – remove shells and tails
6 tbsp oil
1 tbsp sugar
salt to taste
1 big onion – thinly sliced
2 red chillies)
1 green chilli) thinly sliced
1½ cups thick coconut milk – extracted from 230 g grated coconut mixed with 1 cup water
1 stalk lemon grass (*serai*) – bruise
1 small piece *lengkuas* – bruise
½ teacup light tamarind juice – made with 2 tsp tamarind

Grind to a smooth paste:

10 dried chillies – soak to soften and remove seeds for a mild flavour
2 candlenuts (*buah keras*)
1 tsp dried shrimp paste (*blacan*)
6–8 shallots

Method:

Heat oil in deep frying pan and fry ground ingredients together with *serai* and *lengkuas.* Add salt and sugar and fry till fragrant then add liver and shrimps. Stir fry, mixing evenly with the fried paste, for a few minutes, then add thick coconut milk together with tamarind juice. Reduce heat and leave to simmer till gravy is fairly thick, then add sliced chillies and onions and cook till almost dry.

Potato Cutlets

Ingredients:

2 large potatoes – peel and cut into thick slices
5–6 shallots – slice thinly
2 spring onions – chop finely
½ tsp pepper
salt to taste
1 egg – beat lightly
oil for deep frying

Method:

Heat oil in frying pan and fry potatoes till golden brown, then put in a bowl. Fry shallots till brown and add to the potatoes. While potatoes are still warm, add pepper, spring onions and salt and mash till smooth. Shape into small ovals or rounds and coat with beaten egg. Deep fry till golden brown.

Sambal Terong

Ingredients:

2 green eggplant (*brinjal* or *terong*) – cut into half, season with salt and leave for 5 minutes
8 fresh chillies
1 piece dried shrimp paste (*blacan*) (5cm x 2.5 cm or 2 in x 1 in) – flatten to make a patty and roast over low heat till brown
1 tsp sugar
salt to taste

Method:

Grill *brinjal* for 10–12 minutes till flesh is soft. Scrape flesh from the skin with a spoon and put in a bowl. Discard skin.
Grind chillies and *blacan* and add to pulp together with sugar and salt, and mix well.

Urap Taugeh
(Bean Sprouts with Grated Coconut)

Ingredients:
300 g (½ kt) bean sprouts (*taugeh*) – remove tails, wash and drain
1 cup grated coconut
salt to taste
2 limes – extract juice

Grind to a fine paste:
10 dried chillies – soak in water to soften
5 shallots
15 dried prawns – soak, wash and drain

Method:
Place a shallow frying pan over low heat.
Put grated coconut into the dry frying pan and add the ground ingredients and salt. Keep on stirring until fragrant. Add bean sprouts and stir for 1–2 minutes only. Do not overcook the bean sprouts.
Remove from fire and sprinkle lime juice over the bean sprouts. Mix before putting in serving dish.

Serundeng
(Spiced Grated Coconut)

Ingredients:
1 cup grated coconut (with skin)
½ teacup tamarind juice – made with 2 tsp tamarind
1 tbsp sugar
1 tsp salt
2 fragrant lime leaves (*daun limau perut*)
1 tbsp meat curry powder (see pg. 6)

Grind into a smooth paste:
3–4 shallots
1 stalk lemon grass (*serai*)
1 small piccc *lengkuas* (1 cm or ½ in thick)
1 small clove garlic

Method:
Put the grated coconut in a shallow saucepan and add ground ingredients, curry powder, tamarind juice, sugar, lime leaves and salt. Mix well and cook over low heat, turning the contents over from time to time to prevent burning, until dry and crumbly.

Tempe
(Fermented Soya Bean Cake)

Ingredients:
2 packets fermented soya bean cake (*tempe*) – use whole
1 tsp turmeric powder
salt to taste
2 tsp water
oil for deep frying

Method:
Mix salt, turmeric powder and water into a paste and rub it into the pieces of *tempe*.
Heat oil and fry *tempe* till brown.

Ayam Panggang Gahru

Ingredients:
1 chicken, weighing approximately 1.5 kg (2½ kt) – clean chicken, use whole or cut into 4 pieces
450 g (¾ kt) grated coconut – add 2 cups water and extract 2 cups milk
4 tbsp grated coconut (to be taken from the 450 g) – roast over low heat till golden brown, then pound or grind into a smooth paste
1 fragrant lime leaf (*daun limau perut*)
2 tbsp coriander (*ketumbar*) powder
1 tsp cumin (*jintan puteh*) powder
1 stalk lemon grass (*serai*) – bruise
2 small limes (*limau kesturi*) – extract juice
6 tbsp cooking oil

Grind to a fine paste:
6 fresh chillies – seed for mild flavour
12 dried chillies – seed for mild flavour and soak to soften
4 candlenuts (*buah keras*)
5 shallots
1 small almond-sized piece ginger

Method:
Rub roasted, ground coconut into the chicken evenly, then rub coriander and cumin powder, ground ingredients and salt. Add lime leaf and *serai*.
Heat oil in a deep pot then put in the whole chicken with all the spices and fry till fragrant. Add coconut

milk and simmer. When gravy is fairly thick, remove chicken from heat, put in a tray and grill for 15–20 minutes. Remove from grill and sprinkle chicken with lime juice.

NANGKA LEMAK
(Jack Fruit in Coconut Milk)

Ingredients:

450 g (¾ kt) young unripe jackfruit (*nangka*) – peel and cut into slices about 3 cm (1¼ in) thick (oil knife before cutting)
450 g (¾ kt) grated coconut – first add 2 cups water and extract equal amount of milk (thick milk): add another 2 cups of water and extract equal amount of milk (thin milk)
120 g (3 tah) salted *ikan kurau* bones, or dried salted cod – wash
150 g (¼ kt) shrimps – peel
4 sprigs basil (*daun selaseh*), optional – break each into 3 pieces
salt to taste

Grind coarsely:

2 tsp small white dried whitebait (*ikan bilis*) – soak beforehand
5 shallots
½ tsp black peppercorns
2 dried chillies – soak to soften
1 thin slice fresh turmeric

Method:

Boil *nangka* in salted water for about 10 minutes or until half cooked. Drain and put in a large pot with salt fish bones, shrimps, salt, *daun selaseh*, ground ingredients and thin coconut milk.
Bring to the boil over medium heat, stirring while it is cooking. Boil for a minute or two then add thick coconut milk. Bring back to the boil over low heat, and cook stirring all the time for 10–15 minutes.

Note:

450 g (¾ kt) of *nangka* is approximately one quarter of a small fruit.

1. *Green Beans & Tapioca Leaves* **2.** *Salted Unripe Mango (Sidedish)* **3.** *Sambal Asam* **4.** *Ikan Panggang* **5.** *Nangka Lemak*

SAMBAL SARDINE BLIMBING

Ingredients:
1 can local sardines (15 oz or 425g) (see glossary)
Grind to a smooth paste:
  12–15 dried chillies – soak beforehand
  10 shallots
  1 thin slice ginger
  1 clove garlic
  4 candlenuts (*buah keras*)
3 tbsp oil
10–12 *blimbing* – cut in half lengthwise
2 large onions – cut into thick slices
salt to taste

Method:
Heat oil, add ground ingredients and salt and fry till fragrant. Add sardines (include sauce, if any) and stir, mixing well with paste. After a minute or two, add onions and *blimbing*; stir fry for about 1 minute then remove from heat.

SARDINES WITH PINEAPPLE IN COCONUT MILK

Ingredients:
1 can local sardines (15 oz or 425 g) (see glossary)
Grind to a smooth paste:
  10 dried chillies – soak to soften
  1 clove garlic
  1 piece ginger – size of almond
  1 big onion
  1 cm (½ in) cube fresh turmeric
1 stalk lemon grass (*serai*) – bruise
½ small pineapple – cut into small wedges
230 g (6 tah) grated coconut – add 1½ cups water and extract equal amount of milk
salt to taste
5 tbsp oil

Method:
Heat oil, add ground ingredients and *serai* and fry till fragrant. Add pineapple, coconut milk and salt and bring to the boil, stirring now and then. When boiling, add sardines and leave to cook for 5 minutes.

1. *Young Corn & Mushroom in Coconut Milk*
2. *Sardine Salad*
3. *Sardines with Pineapple in Coconut Milk*
4. *Sambal Sardine Blimbing*

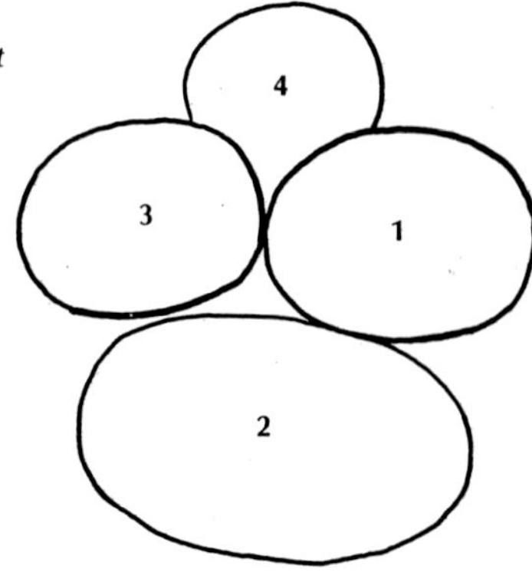

AYAM BRAND
SARDINE
AYAM BRAND
SARDINE

SARDINE SALAD

Ingredients:

1 can local sardines (15 oz or 425 g) (see glossary)

Cut into small squares:

½ cucumber – peel
¼ pineapple
2 medium sized tomatoes

1 big onion – chop
2 red chillies) slice
1 green chilli)
pinch of salt
pepper
some small coriander leaves for garnishing – chop

Method:

Arrange sardines on a platter. Mix all the other ingredients together and spread over sardines. Sprinkle with coriander leaves before serving.

URAP PISANG
(Sliced Banana in Grated Coconut)

Ingredients:

12 ripe *pisang kepok* or *pisang nipah* – wash unpeeled, and steam for 15 minutes
230 g (6 tah) grated coconut – remove skin before grating
2 tbsp sugar
½ tsp salt

Method:

Peel steamed bananas and cut into slices. Mix coconut with sugar and salt, add banana slices and mix.

SAMBAL ASAM

Ingredients:

1 piece dried shrimp paste (*blacan*) (2 x 2 x 1 cm or ¾ x ¾ x ½ in) – roast in dry pan
9 dried chillies – soak in warm water
½ tbsp tamarind mixed with ½ cup water-strain juice
½ tsp salt
2 tsp sugar

Method:

Pound chillies and *blacan* to a fine paste, then mix with tamarind juice. Add salt and sugar and stir well.
This sauce may also be served with boiled vegetables – e.g. long beans, cabbage, *kangkong* (water convolvulus) or ladies fingers (okra).

IKAN PANGGANG
(Grilled Fish)

Ingredients:

Any medium to large sized fish may be used – e.g. *Ikan Kembong, Ikan Terubok, Ikan Parang, Ikan Selar Kuning* (mullet, bream, snapper etc.)
Rectangular pieces of banana leaves large enough to wrap fish

Method:

Clean the fish, make 2 diagonal cuts on each side and rub with salt. Wrap each fish in a banana leaf and grill under a gas or electric griller. Alternatively, cook over an open fire.
Serve with *Sambal Asam.*

GREEN BEANS AND TAPIOCA LEAVES
IN COCONUT MILK

Ingredients:

2 bundles young tapioca leaves
60 g (1½ tah) dried green beans – soak overnight
2 red chillies
4 shallots
2 heaped tbsp small dried whitebait (*ikan bilis*) – soak in water for a few minutes

Method:

Boil tapioca leaves in water to which a little salt has been added. When leaves are tender, cool then cut into 6 cm (2½ in) lengths.
Boil dried green beans in water sufficient to cover them till they are cooked. Do not discard water. Pound or grind chillies and shallots to a fine paste then add *ikan*

bilis and 230 g (6 tah) grated coconut – add 3 cups water and extract equal amount of milk.
salt to taste

Note:
When buying dried green beans, choose the big variety.

pound coarsely.
In a deep pot, put the chopped tapioca leaves, cooked green beans and the little cooking water that is left, ground ingredients, coconut milk and salt. Bring to the boil and cook over low heat for another 10 minutes, stirring frequently.
Serve with a side dish of unripe mango to which salt and birds-eye chilli *(chilli padi)* have been added.

STEAMED IKAN BILIS FRITTERS

Ingredients:
230 g (6 tah) plain flour
120 g (3 tah) rice flour
150 g (¼ kt) steamed dried whitebait (*ikan bilis*)
3 eggs
2 tsp coriander (*ketumbar*) seeds) roast then grind
¼ tsp cumin *(jintan puteh)* seeds) coarsely
1 small bunch big coriander leaves – chop fine
salt to taste
200 ml water
oil for deep frying

Method:
Put the two types of flour, ground spices and salt in a bowl. Break eggs into flour, add coriander leaves and water, stir and mix well.
Add fish and mix.
Heat oil then drop mixture by the spoonful into oil using a dessertspoon. Fry till golden brown.
Serve with *Sambal Cuka.*

OTAK-OTAK
(Steamed Spicy Fish in Banana Leaves)

Ingredients:
450 g (¾ kt) *ikan tenggiri* (Spanish mackerel or bonito) – buy tail end)
Pound or grind:
1 stalk lemon grass *(serai)*
1 piece *lengkuas* (size of walnut)
1 small clove garlic
10 shallots
3 heaped tbsp grated coconut
salt to taste
3 heaped tbsp fish curry powder (see pg. 6) – mix with sufficient water to form a paste
5 rectangular pieces banana leaves – each measuring 25 x 15 cm (10 x 6 in)

Method:
Clean fish, remove scales then fillet the fish. Using a spoon, scrape off flesh and chop.
Combine ground ingredients and salt with curry paste then add fish and mix well.
Put 2 tablespoons of mixture in centre of banana leaf and fold, bringing the two long edges towards the centre so that they overlap. Secure ends with toothpicks. Steam for about ½ hour.
Makes about 5.

Note:
The skin and bones of the fish can be used for making fish stock.

YOUNG CORN AND MUSHROOM IN COCONUT MILK

Ingredients:
1 can (225 g) young corn –use whole
1 small can (198 g) button mushroom
300 g (½ kt) shrimps – peel
Grind to a fine paste:
6 shallots
3 candlenuts (*buah keras*)
1 piece fresh turmeric – (1 cm or ½ in cube)
3 red chillies
4 tbsp oil
230 g (6 tah) grated coconut – add 1½ cups water and extract equal amount of milk
salt to taste
20 birds-eye chilli (*chilli padi*)

Method:
Heat oil, fry ground ingredients till fragrant. Add corn and mushroom, stir for a minute, then add coconut milk and salt and bring to the boil. Add *chilli padi,* leave to boil for a little longer then add shrimps and continue to boil, stirring all the time, for another 10 minutes.

SAYOR KANGKONG KELEDEK (Water Convolvulus and Sweet Potatoes in Coconut Gravy)

Ingredients:
450 g (¾ kt) grated coconut – mix with 4 cups water and extract equal amount of milk
300 g (½ kt) sweet potatoes – peel and cut into medium sized pieces
salt to taste
120 g (3 tah) shrimps – peel
600 g (1 kt) *kangkong* (water convolvulus)–wash and cut into 5 cm (2 in) lengths; discard fibrous end of stalk

Pound or grind fine:
2 red chillies
6–8 shallots
½ tsp dried shrimp paste (*blacan*)

Method:
In a pot, combine ground ingredients with 3 cups coconut milk, salt and sweet potatoes. Bring to the boil; when coconut milk is about to boil, keep stirring to prevent it from curdling.
When potatoes are a little soft, add shrimps, *kangkong* and the remaining cup of coconut milk. Bring to the boil, stirring all the time, then cook for 5 minutes before removing from heat.

FRIED KANGKONG WITH BLACAN

Ingredients:
600 g (1 kt) *kangkong* (water convolvulus) – cut into 10 cm (4 in) lengths and shred stalks
150 g (¼ kt) shrimps – peel
salt to taste
4 tbsp oil

Grind to a fine paste:
1 piece dried shrimp paste (*blacan*) – (2x2x1 cm or ¾x¾x½ in)
2 cloves garlic
8 dried chillies – soak beforehand
8 shallots – peel

Method:
Heat oil, add ground ingredients and salt. Fry till fragrant then add shrimps. Stir, add *kangkong* stalks, stir, then add leaves. Keep stirring till vegetables are cooked. Do not overcook.

1. *Onion and Chilli Salad*
2. *Asam Pedas Melaka*
3. *Urap Pisang*
4. *Otak Otak*
5. *Sayor Kangkong Keledek*

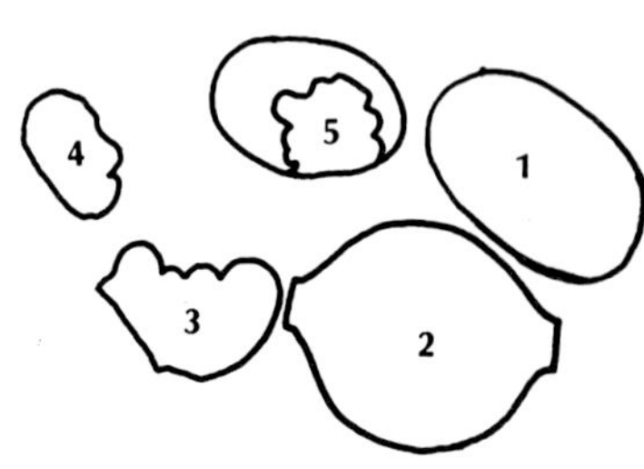

ASAM PEDAS MELAKA (KUAH LADA)

Ingredients:

1 tbsp coriander *(ketumbar)* seeds)
½ tsp cumin *(jintan puteh)* seeds) roast in dry
½ tsp fennel *(jintan manis)* seeds) pan
1 piece fresh turmeric (size of almond)
25 dried chillies – soak
1 tbsp dried shrimp paste (*blacan*)
2 tbsp grated coconut – roast in dry pan over low heat
2 cloves garlic
1 small piece ginger (size of almond)
1 onion
600 g (1 kt) *ikan tenggiri* (Spanish mackerel or bonito) – cut into 6–8 pieces about 1 cm (½ in) thick; rub with salt then wash
¼ medium sized cabbage – cut into 5 wedge-like slices; cut off hard stem
10 tbsp oil
1½ tbsp tamarind – add 3 cups water, mix and strain to get tamarind juice
salt to taste

Method:

Grind the following ingredients starting with coriander then cumin, fennel, turmeric, chillies, *blacan*, coconut, garlic, ginger and onion, adding each ingredient only when the previous one is finely ground.
Heat oil, add ground ingredients and fry till fragrant. Add salt and tamarind juice, stir, then add fish. Cover and bring to the boil then add cabbage slices. As soon as cabbage is cooked, remove from heat. Do not overcook cabbage.

CHICKEN CURRY (INDONESIAN STYLE)

Ingredients:

1 chicken weighing about 1.8 kg (3 kt) – cut into large pieces

Grind to a fine powder:

1½ tsp peppercorns
3 slightly heaped tbsp coriander (*ketumbar*) seeds – roast in dry pan before grinding

Grind to a smooth paste:

12–15 shallots – peel
3 red chillies
2 cloves garlic
6 thin slices ginger

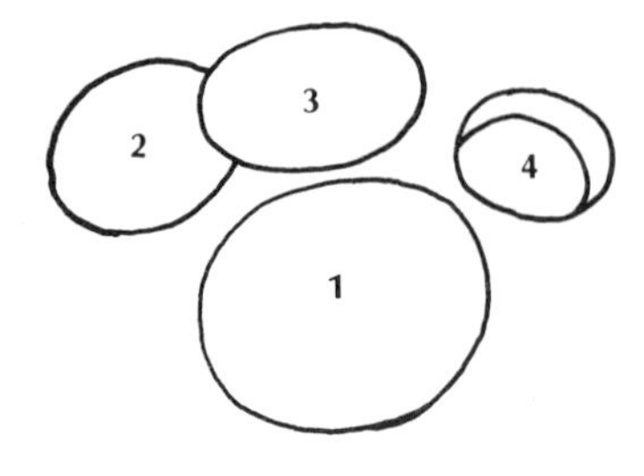

1. *Steamed Nasi Uduk, Fried Kangkong, Sambal Shrimps with Quail Eggs*
2. *Omelette*
3. *Chicken Curry (Indonesian Style)*
4. *Penggat Pisang*

1 stalk lemon grass (*serai*)) bruise
1 piece *lengkuas* (size of walnut)) bruise
¼ tsp turmeric powder
salt to taste
230 g (6 tah) grated coconut – mix with 2 cups water and extract equal amount of milk
1 cup water
8 tbsp oil

Method:
In a large bowl, combine chicken with all the ground ingredients, *serai, lengkuas,* turmeric powder, and salt.
Heat oil, add chicken and fry till fragrant. Add water and coconut milk, then cover and bring to the boil. Simmer till chicken is tender, stirring now and then during cooking. The consistency of the gravy should be a little thick.

SAMBAL SHRIMPS WITH QUAIL EGGS

Ingredients:
600 g (1 kt) shrimps – peel and devein
20 quail eggs (or 4 hen eggs) – boil then remove shell
1 tbsp tamarind – mix with 4 tbsp water and strain to get juice
½ cup water
1 tbsp sugar
salt to taste
6 tbsp oil

Grind to a fine paste:
25 dried chillies – soak beforehand
1 piece dried shrimp paste (*blacan*) (2x2x1 cm or ¾x¾x½ in)
2 medium sized onions
1 clove garlic

Method:
Heat oil, add ground ingredients and fry till fragrant. Add sugar and salt, stir, add tamarind juice and water, then continue cooking till gravy is fairly thick. Add shrimps and stir. When shrimps are cooked, add eggs, stir and remove from heat.
Do not overcook eggs.

IKAN BAWAL CUKA

Ingredients:
salt to taste
½ tbsp sugar
1 tbsp local vinegar (see glossary)
100 ml water
8 tbsp oil
1 *ikan bawal* (pomfret, flounder or John Dory) weighing about 900 g (1½ kt) – clean well, make two slashes on each side of the fish and rub with salt.
2 red chillies) cut each into two lengthwise
2 green chillies) cut each into two lengthwise
2 big onions – slice rather thick
6 medium sized tomatoes – slice rather thick

Grind to a fine paste:
10 red chillies
1 onion
1 clove garlic
6 candle nuts (*buah keras*)

Method:
Deep fry fish till brown, then put aside on a dish. Heat 8 tablespoons oil in a separate pot and fry ground ingredients till fragrant. Add salt, sugar, vinegar and cook for a while. Add water, tomatoes, onions and chillies. Cook for a few minutes then pour over fish.

BITTER GOURD WITH TAUKWA

Ingredients:
1 large bitter gourd – cut in half then cut each half into slices about 0.5 cm (¼ in) thick; sprinkle with salt and let stand for about 15 mins; wash
300 g (½ kt) shrimps – peel
2 large pieces hard beancurd (*taukwa*) – cut into squares and deep fry
1 cup grated coconut – add 1 cup water and extract 1 cup milk
salt to taste
4 tbsp oil

Grind to a fine paste:
10 dried chillies – soak to soften
1 medium sized onion
1 clove garlic

Method:
Heat oil, add ground ingredients and salt and fry till fragrant. Add shrimps and fry for a few minutes. Add *taukwa* and coconut milk, and bring to the boil. When gravy thickens, add bitter gourd and cook till fairly dry, stirring now and then.

PENGGAT PISANG
(Banana in Gula Melaka Syrup)

Ingredients:

600 g (1 kt) palm sugar (*gula Melaka*)
3 cups water
5 tbsp granulated sugar
10 ripe bananas (*pisang kepok*) – cut each in two, lengthwise
1 fragrant screwpine leaf (*daun pandan*) – wash and tie
3 small bundles bean thread vermicelli (*sohoon*) – soak for a few minutes then cut into shorter lengths

Method:

Combine *gula Melaka*, granulated sugar, water and *pandan* leaf in a saucepan and put over heat to dissolve sugar.
When sugar has dissolved, strain syrup into another saucepan, add *sohoon* and bananas. Bring to the boil and leave to cook for 5 minutes.
Serve with Coconut Milk Sauce.

COCONUT MILK SAUCE

Mix 230 g (6 tah) grated coconut (remove skin) with ½ cup boiled water and extract the thick coconut milk, then add a pinch of salt.

OMELETTE

Ingredients:

4 eggs
1 big onion – chop
1–2 red chillies – slice
1 spring onion) chop
2 stalks big coriander leaves) chop
½ tsp pepper
salt to taste
1 tbsp oil

Method:

Break eggs into bowl, beat, add chopped and sliced ingredients, pepper and salt and mix well.
Heat oil. Pour half the egg mixture into pan and fry over moderate heat, first one side then the other.
Repeat, using up mixture.

Alternative Method:

Heat oil. Pour egg mixture into rings for poached eggs to make little omelettes.

STEAMED NASI UDUK

Ingredients:

600 g (1 kt) rice – wash and soak overnight in water to which a teaspoon of salt has been added
450 g (¾ kt) grated coconut – remove skin before grating; add 3½ cups water and extract an equal amount of milk
3–4 fragrant screwpine leaves (*daun pandan*)

Note:

To serve, put the cooked Nasi Uduk on a platter, sprinkle shredded omelette over it and garnish with cucumber slices. If making little omelettes, cut each in half and arrange them around the platter of rice.

Method:

Boil water in steamer.
Drain rice, put in a steaming tray, pour 2 cups coconut milk and break *pandan* leaves over it. Steam for 15–20 minutes, then uncover, stir rice, add 1½ cups coconut milk, stir and continue steaming. After 15 minutes, stir rice again and continue cooking for another 15 minutes.
Sufficient for 6–8 people.

SAMBAL CUKA

Ingredients:

10 red chillies
1 thick slice ginger (about 1 cm or ½ in thick)
1 clove garlic
2 tbsp local vinegar – (see glossary)
1 tbsp sugar (slightly heaped)
1 tsp salt

Method:

Pound chillies, ginger and garlic until fine, then put in a bowl and mix with vinegar, salt and sugar.

PAPAYA WITH SHELLFISH

Ingredients:

600 g (1 kt) cockles or clams (*kerang or seehum*) — scrub thoroughly
2 tsp small white dried whitebait (*ikan bilis*)
½ tsp black peppercorns) grind fine
5 shallots)
salt to taste
1 small unripe papaya (about 450 g or ¾ kt) — peel and cut into slices
2½ cups water

Method:

Put all the above ingredients in a pot. Bring to the boil. Boil for about 5 minutes before removing from heat.

IKAN MASAK KUNING

Ingredients:

1 small *ikan tenggiri* (Spanish mackerel or bonito) — cut into about eight pieces
230 g (6 tah) grated coconut — add 2 cups water and extract 2 cups milk
1 stalk lemon grass (*serai*) — bruised
20 birds-eye chillies (*chilli padi*)
10–12 *blimbing* — cut in half
salt to taste

Grind to a fine paste:

1 piece fresh turmeric — size of almond
8 shallots
4 dried chillies — soak to soften

Method:

Put coconut milk, fish, *serai*, ground ingredients, salt and *chilli padi* in a cooking pot. Put over low heat and bring to the boil slowly, stirring now and then. Add *blimbing* and boil for 5 minutes.

1. *Steamed Ikan Bilis Fritters* 2. *Ikan Bawal Cuka*
3. *Papaya with Shellfish*

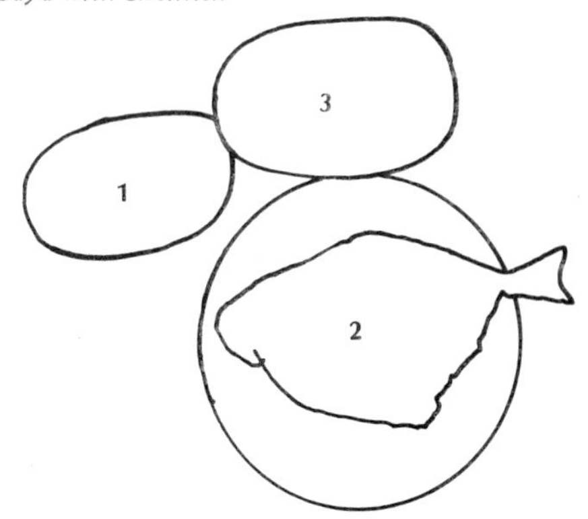

SAMBAL SOTONG KERING

Ingredients:

2 large pieces dried cuttlefish (*sotong*) – cut into small pieces and soak in water for 1–2 hours
1 tbsp tamarind – add ½ cup water and extract ½ cup juice
1 tbsp sugar
salt to taste
5 tbsp oil

Grind to a fine paste:

2 medium sized onions
20 dried chillies – soak and seed
1 piece dried shrimp paste (*blacan*) (2.5 cm or 1 in square)

Method:

Heat oil, add ground ingredients, salt and sugar, and fry till fragrant. Add *sotong,* fry for a while then add tamarind juice and leave to cook till fairly dry.

SAMBAL BLACAN

Ingredients:

1 piece dried shrimp paste (*blacan*) – (2x2.5x1 cm or ¾x1x½ in) – roast in a dry pan
8 fresh red chillies – cut each in three
salt to taste
2 small limes (*limau kesturi*)

Method:

Put salt in pounder with a couple of chillies and pound, gradually add more chillies till all are used up. Finally, add *blacan* and continue to pound.
The final mixture should be neither too fine nor too coarse.
Squeeze lime juice over paste when serving.

1. *Sambal Blacan* **2.** *Sambal Sotong Kering* **3.** *Ikan Masak Kuning* **4.** *Bitter Gourd with Taukwa*

North Indian Dishes

CHAPATI

Ingredients:
450 g (¾ kt) fine wholemeal flour (*atta*)
2 tbsp cooking oil
1 cup water mixed with 1 tsp salt

Method:
Sieve flour into a bowl. Make a hollow in the centre and pour oil into it. Rub oil into flour then add half the amount of water. Knead well for 10 minutes to form a soft dough.
Add the rest of the water gradually. Knead for another 10 minutes then cover with damp cloth and leave to stand for 45 minutes to 1 hour. Divide dough into balls the size of walnuts, and roll each on lightly floured board into thin, round pancakes.
Heat griddle (*tawa*) or a heavy-based frying pan. When very hot, put one chapati on it. When small blisters appear on the surface, cook other side. Leave for a second then press lightly with a folded tea towel so that bubbles appear and when lightly brown, remove and wrap in a tea towel till ready to serve.
Brush a little *ghee* or oil on one side before serving. (Optional)

Note:
Leaving the dough to stand before rolling, and pressing chapatis with a tea towel while cooking makes the chapatis light.

NORTH INDIAN ALU
(Potato Curry)

Ingredients:
600 g (1 kt) potatoes – scrub well, boil and cut each into 6 wedges; do not peel
Roast in dry pan and grind coarsely:
1 tbsp coriander (*ketumbar*) seeds
6 dried chillies
1/2 tsp turmeric powder
1 large tomato – chop coarsely
1 medium sized onion – grate coarsely
1 cup water
salt to taste
6 tbsp oil

Method:
Heat oil, add potatoes and stir fry. Add ground ingredients and turmeric powder, stir, add onion, tomato and water, stir, add salt and leave to cook till gravy is quite thick.

1. *North Indian Chicken Curry*
2. *North Indian Alu (Potato Curry)*
3. *Kheema (Minced Meat Curry)*
4. *Chapati*

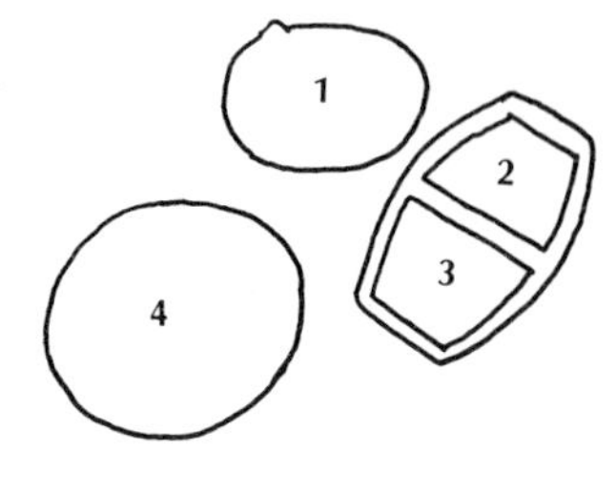

KITCHREE RICE

Ingredients:

600 g (1 kt) Basmati rice – wash and drain
170 g (4½ tah) yellow lentils (*dhal*) – soak for 2 hours, then wash and drain
1 medium sized onion – slice thin
1 stick cinnamon 5 cm (2 in) long
2 cardamoms
4 cloves
2 fragrant screwpine leaves (*daun pandan*) – wash and tie
3 slightly heaped tbsp *ghee*
5 small cloves garlic) grind to a
1 slice ginger (1 cm or ½ in thick)) smooth paste
water – equal in volume to amount of rice used
1 tbsp plain yoghurt
1 small tin evaporated milk (170 g)
½ tsp yellow food coloring
salt to taste

Method:

Combine rice with *dhal* in a bowl.
Heat *ghee,* fry onion, cinnamon, cardamoms, cloves, *pandan* leaves and garlic-ginger paste till fragrant.
Add water, yoghurt, milk, salt and colouring and bring to the boil. As soon as it boils, add rice and *dhal,* stir, cover and leave to boil. Lower heat when amount of water is reduced and continue cooking till water is absorbed, then stir rice and continue cooking over very low heat, keeping pot covered.
When steam emits from pot, rice is cooked. If preferred, colouring may be added when rice is almost cooked.

KHEEMA
(Minced Meat Curry)

Ingredients:

230 g (6 tah) minced beef or mutton
Grind to a fine paste:
- 4 slices ginger (each 1 cm thick)
- 3 cloves garlic

3 tbsp meat curry powder (see pg. 6)
2 cups water
salt to taste
1 small bunch big coriander leaves – chop
120 g (3 tah) green peas
1 green chilli – slice thinly
Cut into small cubes:
- 2 medium sized potatoes – peel
- 2 medium sized tomatoes
- 1 big onion

Ingredients for frying:
- ½ onion – slice thinly
- 3 cardamoms
- 4 cloves
- 1 stick cinnamon – 4 cm (1½ in) long

6 tbsp oil

Method:

In a bowl, combine minced meat with ginger-garlic paste, curry powder, salt and ½ cup water from allowance.
Heat oil, add ingredients for frying. When onions are brown, add minced meat and fry till fragrant. Add potatoes, stir, then add 1½ cups water and bring to the boil. Cover and continue cooking till potatoes are soft, stirring now and then. Add tomatoes, onion, peas and chilli, cover and cook for 3–5 minutes.
When serving, sprinkle with chopped coriander leaves.

COCONUT SAMBAL

Ingredients:

1 cup grated coconut
5 shallots – peel
2 green chillies
1½ tsp tamarind – remove seeds
1 cm (½ in) cube ginger
1 clove garlic
1 sprig curry leaves (*karuvapillai*) – discard stalk
½ tsp salt

Method:

Grind coconut, then add chillies, ginger, garlic, salt, tamarind and shallots. When these have been ground to a fine paste, put into a serving dish.

NORTH INDIAN CHICKEN CURRY

Ingredients:

1 chicken weighing about 1.75 kg (3 kt) – cut into fairly large pieces and remove skin
1 tsp turmeric powder
2 tbsp plain yoghurt
1¾ cups water
salt to taste
3 onions – grate
8 tbsp oil

Roast in dry pan till lightly brown:

4 level tbsp coriander (*ketumbar*) seeds
1 tsp cumin *(jintan puteh)* seeds
18 dried chillies – break each into 3 pieces

Method:

Grind roasted ingredients. In a bowl, combine chicken with ground spices, turmeric powder, salt, yoghurt and ¼ cup water from allowance.
Heat oil, add chicken, fry till fragrant, then add onion. Stir fry for a while then add 1½ cups water, cover and cook over low heat.
When chicken is tender and gravy a little thick, remove from heat.

MUTTON CHOPS WITH GRAVY

Ingredients:

600 g (1 kt) mutton or lamb chops (about 8 chops)

Grind to a fine paste:

6 green chillies
4 red chillies
3 slices ginger (each 1 cm or ½ in thick)
5 cloves garlic

2 tbsp plain yoghurt or juice of 1 lime
2 tbsp coriander (*ketumbar*) powder
1 tbsp chilli powder
2 onions – chop
4 tomatoes – chop
6 stalks big coriander leaves – chop
salt to taste

Ingredients for frying:

1 small onion – slice thinly
4 cardamoms
5 cloves
1 stick cinnamon 5 cm (2 in) long

10 tbsp oil
4 cups water

Method:

In a large bowl, combine chops with all the above ingredients except ingredients for frying, oil and water.
Heat oil, add ingredients for frying. When fragrant, add mutton and fry till fragrant then add water. Cover and boil, removing cover and stirring from time to time.
Continue cooking till mutton is tender and the gravy is very thick.

PILAU RICE WITH CASHEW NUTS

Ingredients:

600 g (1 kt) *Basmati* rice – wash and drain
4 tbsp *ghee*

Ingredients for frying:

1 stick cinnamon 4 cm (1½ in) long
4 cloves
4 cardamoms

Grind to a fine paste:

1 piece ginger – size of a walnut
3 cloves garlic

3 tbsp plain yoghurt
water – equal in volume to the amount of rice used
5 tbsp evaporated milk
salt to taste
1 tbsp rose water
120 g (3 tah) cashew nuts – fry in *ghee* and put aside for garnishing

Method:

Heat *ghee*, add ingredients for frying and ginger-garlic paste. Fry till fragrant, then add yoghurt, water, evaporated milk, salt and rose water and bring to the boil. When boiling, add rice and continue boiling. When rice is drying, reduce heat. Keep pot covered and cook over low heat. When fragrance emits from pot, stir to turn rice over. Cover pot and cook for another 5 minutes, then sprinkle cashew nuts over top.

DRIED MUTTON CHOPS

Ingredients:
450 g (¾ kt) mutton or lamb chops (about 6 pieces)
1 tbsp chilli powder
1 tsp turmeric powder
1 tbsp coriander (*ketumbar*) powder
½ tsp black pepper powder
salt to taste
1 tsp cumin *(jintan puteh)* seeds — soak then crush coarsely
¼ cup water
6 tbsp oil
1 cup water
1 small bunch small coriander leaves for garnishing

Method:
Mix the above spices, salt and ¼ cup water into a paste, and then rub paste over chops.
Heat oil, add chops and fry till fragrant. Add water, cover pan and cook over low heat. When meat is tender and moisture absorbed, remove from heat and drain. Garnish with small coriander leaves and serve hot.

TOMATO PUTCHREE

Ingredients:
1 tbsp cooking oil or *ghee*
6 medium sized tomatoes — cut each into 8 wedges
1 tsp chilli powder
½ tbsp coriander (*ketumbar*) powder
½ tbsp turmeric powder
½ cup water
1 tbsp vinegar
1 tbsp sugar
salt to taste
Ingredients for frying:
1 stick cinnamon — 4 cm (1½ in) long
3 cloves garlic
3 cardamoms
½ onion — slice finely

Method:
Heat oil, add ingredients for frying and fry till fragrant. Add tomatoes and the rest of the ingredients and bring to the boil. Continue boiling for 1—2 minutes, then remove from heat.

1. *Coconut Sambal* **2.** *Mutton Chops with Gravy* **3.** *Dried Mutton Chops* **4.** *Egg Curry with Shrimps & Potatoes* **5.** *Kitchree Rice*

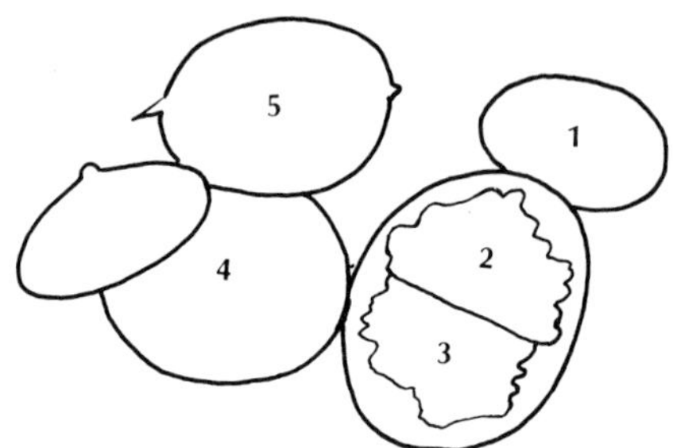

EGG CURRY WITH SHRIMPS AND POTATOES

Ingredients:
5 eggs – boil and shell, then make long, shallow cuts on the surface of each egg
300 g (½ kt) shrimps – peel

Combine in a bowl:
6 tbsp fish curry powder (see pg. 6)
1 tbsp grated coconut – grind fine
½ onion – slice thickly
1 green chilli) slash
1 red chilli) slash
1 sprig curry leaves (*karuvapillai*)
1½ tbsp tamarind – add 3 cups water and strain juice

Ingredients for frying:
1 tsp fish curry spices (see glossary)
1 sprig curry leaves (*karuvapillai*)
½ onion – thinly sliced
6 small potatoes – halve and leave unpeeled
4 tbsp oil

Method:
Heat oil, add potatoes and ingredients for frying. Fry till potatoes and onions are brown, then add ingredients combined in bowl. Cover and bring to the boil. When potatoes are almost cooked, add eggs and shrimps. Continue boiling till potatoes are cooked.

LIME PICKLE

Ingredients:
2 limes (commercially bottled Indian Lime Pickle) – chop
1 large onion – cut in half and slice fine
1 red chilli) slice fine (seed for mild flavour)
2 green chillies) slice fine (seed for mild flavour)

Method:
Combine all the above ingredients in a bowl, mix well and put in serving dish.

1. *Tomato Putchree* **2.** *Tomato Chicken Curry* **3.** *Minced Mutton Balls* **4.** *Fried Potatoes* **5.** *Pilau Rice with Cashew Nuts*

SLICED MUTTON CURRY WITH CABBAGE

Ingredients:

450 g (¾ kt) lean mutton — cut into slices of average size
4 small or 2 large tomatoes — cut into small cubes
1 medium sized onion — chop
2½ tbsp water
2 tbsp coriander (*ketumbar*) powder
3 tsp chilli powder
1 tsp turmeric powder
½ tsp black pepper powder
3 sprigs big coriander leaves — break into pieces
a few mint leaves
10 almonds — scald, remove skin and grind fine
1 stick cinnamon — 4 cm (1½ in) long
2 cardamoms
3 cloves
salt to taste
4 tbsp *ghee* or 8 tbsp cooking oil
2 cups water
230 g (6 tah) cabbage — cut into small pieces

Method:

Combine all the above ingredients **except** cabbage, *ghee* and 2 cups water, in a big bowl.
Heat *ghee* or oil. Put contents of bowl into oil and fry till fragrant. Add water, cover and cook, stirring now and then.
When meat is tender and gravy fairly thick, add cabbage and cook for a few minutes.

MINCED MUTTON BALLS

Ingredients:

230 g (6 tah) minced mutton
2 onions — chop fine
salt to taste
1 tsp pepper
2 heaped tbsp flour
2 small bunches big coriander leaves — chop fine
1 tbsp coriander (*ketumbar*) powder
1 tsp chilli powder
½ tsp turmeric powder
oil for deep frying

Method:

Combine all the ingredients, mix well, then shape into balls.
Heat oil, fry meat balls till golden brown.

NOTE:
Beef or pork may be used instead of mutton.

STUFFED CHICKEN LEGS

For this dish, use chicken thighs and drumsticks only.
Prepare the stuffing first, giving it time to cool.

Ingredients:

2 chicken drumsticks and 2 thighs (approx. 450 g or ¾ kt) — make a slit about 5 cm (2 in) long near the bone of each piece; rub chicken with salt and pepper
oil for deep frying

For Stuffing:

4–5 small florets of cauliflower) chop fine
1 small carrot)
5 French beans — slice very thin
1 tsp coriander (*ketumbar*) powder
½ tsp chilli powder
½ tsp turmeric powder
2½ tbsp water
a few mint leaves — chop fine
salt to taste
1½ tbsp oil

Method:

Heat oil, add vegetables and stir fry. After a minute or two, add coriander, chilli and turmeric powder, mint leaves and salt. Fry, then add water and cook till vegetables are soft and dry. Leave to cool.

To stuff chicken:

Spoon stuffing into cavity of chicken legs and secure with thread.

To cook:

Heat oil then deep fry chicken over low heat.

WATERY DHAL

Ingredients:

1 cup yellow lentils (*dhal*) – soak for about 3 hours, then wash, remove skin and drain
3 cups water
½ onion – slice thick
2 green chillies – slit lengthwise
2 cloves garlic – cut into thick slices
1 stick cinnamon – 3 cm (1¼ in) long
3 cloves
1½ tsp turmeric powder
salt to taste

Method:

Combine the above ingredients in a deep pot; stir to mix, then boil till *dhal* is soft; remove from heat.

Ingredients for frying:

½ onion – slice finely
1 tsp cumin *(jintan puteh)* seeds – to be ground coarsely
3 dried chillies – break each into 3 pieces
2 tbsp *ghee*

Method:

Heat *ghee*, then fry the above ingredients till fragrant and onions brown; pour *ghee* and fried ingredients immediately into cooked *dhal*, cover the pot for a minute then serve.

FRIED POTATOES

Ingredients:

450 g (¾ kt) small potatoes (if large ones are used, cut each into four) – peel and rub salt over them
1 tbsp coriander (*ketumbar*) powder
½ tbsp chilli powder
1 tsp turmeric powder
2½ cups water
6 tbsp *ghee* or cooking oil
big coriander leaves (for garnishing) – chop

Method:

Heat *ghee*, and fry potatoes. When slightly brown, add coriander, chilli and turmeric powder and water. Cover and cook until dry, stirring now and then during cooking.
When serving, garnish with chopped coriander leaves.

TOMATO CHICKEN CURRY

Ingredients:

1.75 kg (3 kt) chicken – cut into medium sized pieces
Grind to a fine paste:
2 pieces ginger – each the size of a walnut
6 cloves garlic
6 medium sized tomatoes – cut into very small cubes
2 medium sized onions – chop fine
4 level tbsp coriander (*ketumbar*) powder
2 level tbsp chilli powder
1 level tsp turmeric powder
5 sprigs big coriander leaves – break into pieces
4 tbsp plain yoghurt
salt to taste
In a big bowl, combine chicken with all the above ingredients.

Ingredients for frying:

1 onion – slice thinly
4 cardamoms
6 cloves
2 sticks cinnamon – each 3 cm (1½ in) long
8 level tbsp *ghee*
3 cups water
2 level tbsp tomato puree
mint leaves – for garnishing

Method:

Heat *ghee*, add ingredients for frying and fry till fragrant. Add chicken, fry for about 5 minutes, then add water, cover pot and bring to the boil.
Mix tomato puree with a little water. When curry is boiling, add tomato puree and continue boiling till chicken is tender.
When serving garnish with mint leaves.

MINCED CHICKEN KEBAB

Ingredients:

230 g (6 tah) chicken meat – mince
1 small bunch big coriander leaves)
1 small bunch mint leaves) chop fine
1 tbsp coriander (*ketumbar*) powder
1 tsp chilli powder
½ tsp turmeric powder
1 tsp white pepper
1½ tbsp golden bread crumbs
salt to taste
oil for deep frying

Method:

Combine all ingredients except oil in a bowl.
Mix well and shape into small ovals. Deep fry till golden brown.
Serve with chilli or mint sauce.

1. *Stuffed Chicken Legs* **2.** *Minced Chicken Kebab* **3.** *Watery Dhal* **4.** *Sliced Mutton Curry with Cabbage*

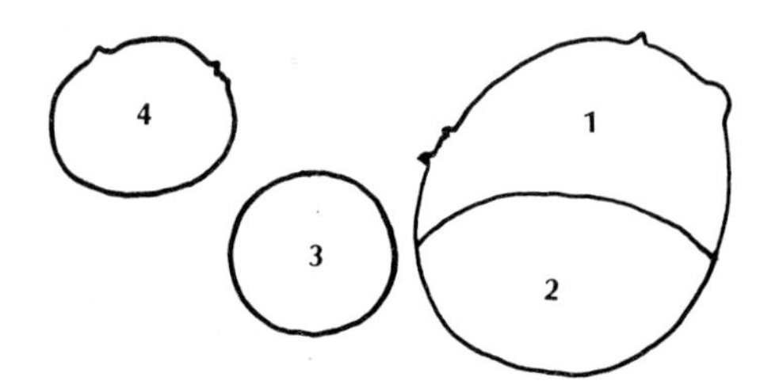

South Indian Dishes

CRAB CURRY

Ingredients:
600 g (1 kt) crabs – cut body into two pieces
1 tbsp coriander (*ketumbar*) powder
1 level tsp cumin *(jintan puteh)* powder
1 tbsp chilli powder
1 tsp turmeric powder
2 sprigs curry leaves (*karuvapillai*)
1 onion – slice thick
salt to taste
1½ cups water
4 tbsp oil
1 tbsp rice – wash and dry roast in dry pan till slightly brown then grind

Method:
Put crabs in a large bowl; add coriander, cumin, chilli and turmeric powder, curry leaves, onion, salt and ½ cup water. Leave to stand for about 10 minutes.
Heat oil, add crabs and keep stirring. When crabs turn red, add 1 cup water and simmer; when boiling, add ground rice to thicken gravy. Continue cooking, stirring all the time. When gravy is slightly thick remove from heat.

FRIED CABBAGE

Ingredients:
350–400 g (9–10 tah) cabbage – shredded
2 green chillies – cut slantwise into fairly big pieces
Ingredients to be fried:
1 big onion – slice thinly
1 sprig curry leaves (*karuvapillai*)
1 tsp brown mustard seeds
1 tsp turmeric powder
1 cup grated coconut – add 1 cup water and extract equal amount of milk
salt to taste
3 tbsp oil

Method:
Heat oil, add ingredients to be fried and chillies, and fry till fragrant. Add cabbage and stir fry for a few minutes. Add coconut milk and salt, and cook over medium heat till gravy boils. Continue cooking until cabbage is just tender then remove from heat.

1. *Fish Head Curry*
2. *Hot Prawn Curry*
3. *Fried Cabbage*
4. *Rasam*
5. *Papadom*
6. *Crab Curry*

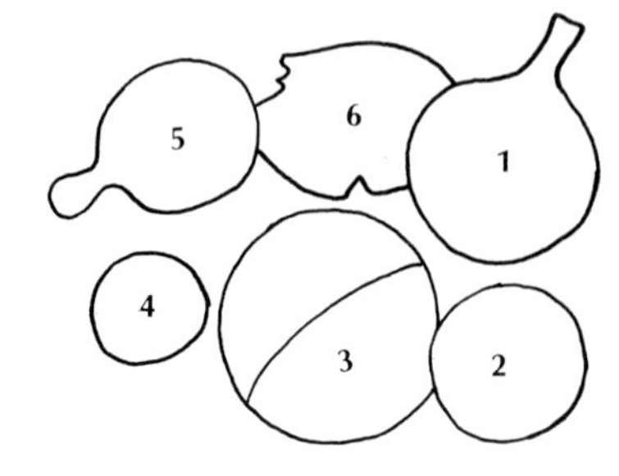

FISH HEAD CURRY

Ingredients:

4 tbsp fish curry powder
salt to taste
1 sprig curry leaves (*karuvapillai*)
1 red chilli
1 green chilli
½ onion – slice thick
1 clove garlic – slice
1 medium sized fresh tomato – to be used whole
½ cup water
2 tbsp tamarind – add 3 cups water, mix and strain to get tamarind juice
8–10 ladies fingers (okra) – slit lengthwise
1 medium sized fish head weighing about 900 g (1½ kt) – rub with salt and wash thoroughly
1 cup grated coconut (optional) – add ½ cup water and extract equal amount of milk

Ingredients to be fried:

2 tsp fish curry spices (*rempah tumis ikan*) (see glossary)
1 sprig curry leaves (*karuvapillai*)
1 medium sized onion – slice thinly

3 tbsp cooking oil

Method:

Put curry powder, salt, curry leaves, chillies, onion, garlic and tomato in a medium sized bowl. Add water. Using a spoon or your hand, coarsely crush the chillies and tomato, then add tamarind juice to get a watery mixture. Set aside.
Heat oil, add ingredients to be fried and fry 1–2 seconds then add ladies fingers and fry till fragrant. Pour in curry mixture and keep pot covered till it is boiling. This is important to prevent the curry from tasting uncooked. Add fish head. If coconut milk is used, add when curry is boiling then bring back to the boil before adding fish head. Boil fish head till it is cooked.

HOT KERALA CHICKEN CURRY

Ingredients:

½ chicken weighing about 750 g (1¼ kt) in weight –remove skin and cut into small pieces
1 piece ginger 3 cm (1¼ in) long – slice fine
½ tsp brown mustard seeds
20 shallots – use whole
1 sprig curry leaves (*karuvapillai*)
salt to taste

Roast in dry pan then grind fine:

12 dried chillies
3 tbsp coriander (*ketumbar*) seeds
½ tsp cumin (*jintan puteh*) seeds
½ tsp fennel (*jintan manis*) seeds

1–2 tsp black peppercorns – grind fine
5 tbsp oil
1 cup water

Method:

Combine chicken with ginger, mustard seeds, shallots, curry leaves, salt and all ground spices. Leave to marinate for about 10 minutes.
Heat oil, add chicken, fry till fragrant. Add water and simmer, keeping the pot covered. Stir once or twice during cooking. When gravy is a little thick, remove from heat.

Note:

For this dish, the best flavour is obtained if the spices are ground on a *batu giling* (grinding stone).

HOT PRAWN CURRY

Ingredients:

600 g (1 kt) big prawns (with shell)
1 level tbsp chilli powder
1 level tbsp coriander (*ketumbar*) powder
1 tsp turmeric powder
2 tsp black pepper powder (reduce amount if milder flavour is required)
2 sprigs curry leaves (*karuvapillai*)
salt to taste
1 cup water
1 big onion – thinly sliced
4 tbsp cooking oil

Method:

Put prawns in bowl. Add chilli, coriander, turmeric, black pepper powder, salt, 1 sprig curry leaves and water and leave for 10 minutes.
Heat oil, add sliced onion and the other sprig of curry leaves and fry till fragrant. Add marinated prawns and fry stirring now and then, until cooked.

INDIAN NASI LEMAK

Ingredients:

600 g (1 kt) *Basmati* rice – wash thoroughly and drain
4 tbsp *ghee*
½ onion – slice thinly
1 stick cinnamon – 4 cm (1½ in) long
2 cardamoms
3 cloves
2 fragrant screwpine leaves (*daun pandan*) – tie into a bundle
450 g (¾ kt) grated coconut – add 3 cups water and extract 3½ cups coconut milk
salt to taste

Method:

Heat *ghee* in a large pot for cooking rice. Add onion, cinnamon, cardamoms, cloves and *pandan* leaves. Fry till onions are brown, then add coconut milk, rice and salt, stir. Cover pot and bring to the boil; when boiling, stir rice.
When rice is beginning to dry, lower heat and continue to cook. Keep the pot covered till the steam escapes, then uncover, turn rice over with a fork and cook for another 5–8 minutes.

Note:
Indian Nasi Lemak, which is popular among South Indian Muslims, should be served with Chicken Curry, Cucumber Salad and Crispy Long Beans.

CUCUMBER SALAD

Ingredients:

2 cucumbers – peel, scrape design with fork on cucumber, and cut into slices of medium thickness
salt to taste
the juice of 3 small limes (*limau kesturi*)
2 red chillies) slice thinly
2 green chillies)
1 onion – cut into half and slice fine
1 cup grated coconut – add ½ cup water and extract ¾ cup milk

Method:

Sprinkle salt over cucumber and leave for 5 minutes, then rinse cucumber and drain.
Put the sliced cucumbers in a bowl, then add lime juice, chillies and onion.
Pour coconut milk dressing over salad.

Note:
If salad is to be kept, boil coconut milk dressing, cool, then pour over salad. If serving immediately, use freshly squeezed coconut milk.

CHICKEN CURRY

Ingredients:

1 chicken weighing about 1.75 kg (3 kt) – wash and cut into medium sized pieces
4 level tbsp meat curry powder (see pg. 6)
salt to taste
½ tsp black pepper powder
1 sprig curry leaves (*karuvapillai*)
6 tbsp oil
2 cups water
6 small potatoes – cut each in half
4 level tablespoons grated coconut – pound till fine

Method:

Combine chicken with curry powder, salt, black pepper, curry leaves and coconut in a large bowl. Add ½ cup water from allowance, then leave to marinate for about 10 minutes.
Heat oil, add chicken and stir fry till fragrant. Add the rest of the water and potatoes, cover pan and boil, stirring now and then. When potatoes are half cooked, reduce heat and let curry simmer. Remove from heat when potatoes are cooked and gravy a little thick.

CRISPY LONG BEANS

Ingredients:

300 g (½ kt) long beans – cut into 3 cm (1¼ in) lengths
1 sprig curry leaves (*karuvapillai*)
1 level tsp turmeric powder
1 level tsp chilli powder
1½ tsp salt
8 tbsp oil

Method:

Put long beans in a bowl, sprinkle with turmeric, chilli powder and salt and mix thoroughly. Leave for 10–15 minutes.
Heat oil, fry curry leaves till crispy then add beans and stir fry from time to time. As beans start to get crisp, keep stirring constantly until the beans are very crisp. Remove from heat.

MUTTON CURRY

Ingredients:
900 g (1½ kt) mutton – cut into medium sized pieces
Grind to a fine paste:
6–8 cloves garlic
3 slices ginger – each 1 cm (½ in) thick
1 red chilli) slash
1 green chilli)
½ medium sized onion – cut into thick slices
½ cup water
2 tbsp plain yoghurt (optional)
2 sprigs big coriander leaves – break into pieces
salt to taste
In a large bowl, combine mutton with all the above ingredients.
Ingredients for frying:
1 stick cinnamon – 3 cm (1¼ in) long
½ medium sized onion – slice fine
4 cloves
3 cardamoms
4 tbsp oil
4 cups water
1½ tbsp tomato puree
5–6 tomatoes – cut into wedges

Curry mixture:
Mix the following with enough water to get a thick mixture:–
4 tbsp coriander (*ketumbar*) powder
2 tbsp chilli powder
½ tbsp cumin *(jintan puteh)* powder
2 tsp turmeric powder

Method:
Heat oil, add ingredients for frying. When fragrant, add mutton and stir fry a little, then add water. Bring to the boil, then simmer until meat is half cooked, then pour in curry mixture. Bring to the boil again, cover pot then simmer. Do not remove lid during this part of the cooking till fragrance emits from the pot.
When gravy is quite thick, add tomato puree mixed with some of the gravy and continue cooking till meat is tender.
When almost cooked, add tomato wedges. Cook for a minute or two then remove from heat.
Serve with cucumber and tomato salad.

1. *South Indian Chicken Curry*
2. *Crispy Long Beans*
3. *South Indian Nasi Lemak*
4. *Fried Mutton Curry*
5. *Cucumber Salad*

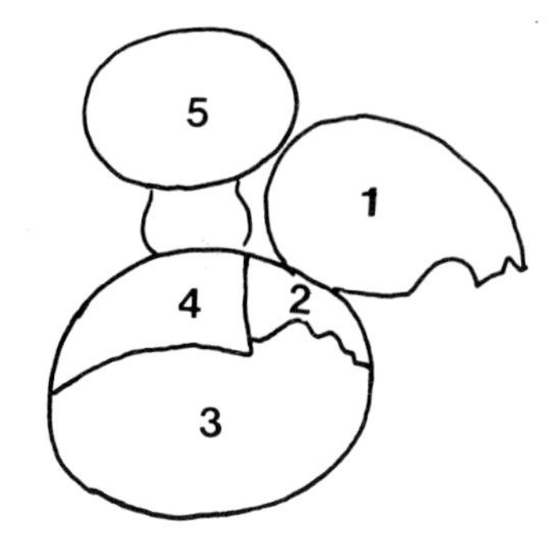

ROTI JALA

Ingredients:

450 g (¾ kt) grated coconut (without skin) – add 3 cups water and extract 4 cups milk
450 g (¾ kt) plain flour – sieve
1 tsp salt
2 eggs
a drop of yellow food colouring (optional)
ghee or cooking oil to grease pan

Method:

Put flour and salt in a bowl. Make a well in the centre. Break eggs into the well and pour in half the coconut milk. Stir, drawing in the flour from the sides. When mixture is smooth, add the rest of the coconut milk and colouring, if used. Stir well so that the batter is well mixed, then strain.
The batter must be neither too thick nor too thin.

To fry:

Heat medium sized frying pan or hot plate. (A non-stick frying pan is ideal.)
When pan is hot, brush it with *ghee* or cooking oil.
When oil is hot, put a ladleful of mixture into Jala cup (see illustration on page 11) and pour mixture into pan in circles till a lacy pancake, the size of the pan, is formed. While pancake is cooking, dab a touch of *ghee* over it.
Remove pancake as soon as it becomes firm.
Repeat till all the mixture is used.

To fold:

To fold the Roti Jala, bring two sides of the roti to meet at the centre. Do the same with the other two sides. Fold once more, so that a small roll is formed. Arrange on a plate.

NOTE:

The coconut milk may be replaced by 1 large tin of evaporated milk mixed with 2½ cups water.

1. *Mutton Curry* **2.** *Roti Jala* **3.** *Chicken Coconut Milk Curry* **4.** *Roti Mariam*

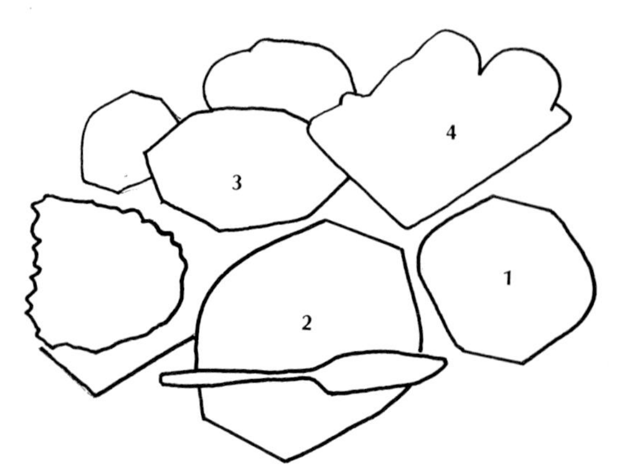

FRIED SPINACH

Ingredients:

300 g (½ kt) spinach – chop off roots, wash and cut fine
450 g (¾ kt) grated coconut – add 1 cup water to get an equal amount of milk

Ingredients for frying:

- ½ onion – slice thinly
- 1 green chilli – cut slantwise into 5–6 pieces
- ½ tsp brown mustard seeds
- ½ tsp turmeric powder
- 1 tsp cumin *(jintan puteh)* powder
- ½ tsp chilli powder
- 3–4 curry leaves (*karuvapillai*)

2 tbsp oil
salt to taste

Method:

Heat oil, add ingredients to be fried. Fry till onions are a light brown then add spinach and stir to mix it with the spices. Add coconut milk and salt and allow to cook, stirring now and then. When spinach is soft and a little gravy is left, remove from heat.

ROTI MARIAM

Ingredients:

450 g (¾ kt) plain flour – sieve
2 eggs
½ cup warm water mixed with 1 tsp salt
2 tbsp *ghee*
3 tbsp cooking oil
oil for deep frying

Method:

Put the flour, eggs, *ghee* and oil into a large bowl. Mix with the hand, then add water, a little at a time, to form a soft dough. Knead dough for about 10 minutes. Shape into lumps the size of a small potato.
On a lightly floured board, roll each lump into a round pancake, about 0.5 cm (½ in) thick.
Deep fry the pancakes.
Serve with chicken or mutton curry.

1. *Fried Spinach* **2.** *Fried Chilli Tairu* **3.** *Salt Fish Curry* **4.** *Sothi*

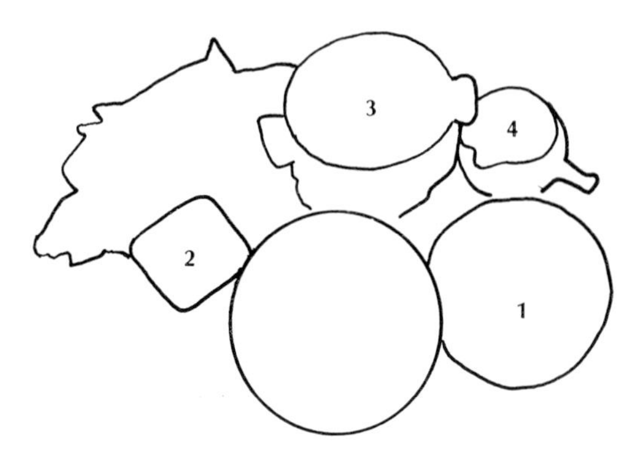

FRIED MUTTON CURRY

Ingredients:
450 g (¾ kt) mutton – cut into small cubes
6 level tbsp meat curry powder (see pg. 6)
Grind to a fine paste:
- 4 cloves garlic
- 1 piece ginger (size of walnut)

salt to taste
8 tbsp oil
Ingredients for frying:
- 1 onion – slice thinly
- 1 stick cinnamon – 4 cm (1½ in) long
- 3 cardamoms
- 4 cloves

1 sprig curry leaves (*karuvapillai*)
2½ cups water
1 red chilli) cut in two lengthwise
1 green chilli)
20 shallots – peel and use whole

Method:
In a large bowl, combine meat with curry powder, ginger-garlic paste and salt and mix well.
Heat oil and cook ingredients for frying till brown.
Add meat and curry leaves, then stir fry till fragrant.
Add water, cover pot and bring to the boil, then simmer, stirring now and then.
When gravy becomes fairly thick and meat is tender, add chillies and shallots. Stir for a few minutes, then remove from heat.

SOTHI

Ingredients:
450 g (¾ kt) grated coconut – add 2 cups water and extract an equal amount of milk
½ tsp turmeric powder
1 green chilli
1 tomato
1 sprig curry leaves (*karuvapillai*)
1 tbsp tamarind – mix with a little of the coconut milk and strain
½ tsp fenugreek (*alba*)
Ingredients for frying:
- ½ onion – slice thinly
- 1 small sprig curry leaves (*karuvapillai*)
- 3 dried chillies – break each into 3 pieces

2 tbsp oil

Method:
Combine coconut milk, turmeric powder, curry leaves, tamarind juice and fenugreek in a pot. Using your hand or a spoon, crush the tomato and the chilli in the coconut milk. Bring to the boil, stirring all the time. When boiling remove from heat.
Heat oil, add ingredients to be fried. When onions are brown, add the fried ingredients to the cooked Sothi and cover for a moment, then remove cover. Stir, then keep covered till ready to serve.

PICKLE

Ingredients:
600 g (1 kt) cucumber – cut into 5 cm (2 in) strips, discarding the pulp
1 large carrot – peel and cut into 5 cm (2 in) strips
30 small round shallots – peel
5 green chillies) slash
5 red chillies)
5 slices ginger, each 1 cm (½ in) thick – shred
20 small cloves garlic
2 tsp chilli powder) add sufficient water
1 tsp turmeric powder) to make a paste
1½ tsp brown mustard seed)
3 tbsp oil
¼ cup water
3 tbsp local vinegar
2 tbsp sugar
salt to taste

Method:
Sprinkle salt generously over cucumber, carrot, ginger, chilli, shallots and garlic and leave in the sun for a few hours, turning vegetables over now and then and get rid of excess moisture.
Heat oil, add chilli-turmeric paste and fry. Add water, vinegar, sugar and salt and stir till sugar dissolves.
Add the vegetables, stir for about 5 minutes then remove from heat. Do not overcook or vegetables will not be crisp.

NASI BRIANI

Ingredients for chicken:

1 chicken weighing approximately 1.5kg (2½kt)— cut chicken into 4 pieces, wash and drain

Grind to a smooth paste:

4 slices ginger – each 1 cm (½ in) thick
4 cloves garlic

1 red chilli)
1 green chilli) slit halfway
4 small tomatoes – slice
3 tbsp (slightly heaped) meat curry powder (see pg. 6)
2 small bunches big coriander leaves – chop coarsely
a small bunch mint leaves
4 tbsp plain yoghurt
1 tsp pepper
salt to taste
1½ cups water
10 tbsp *ghee*

Ingredients for frying:

1 big onion – slice thinly
1 stick cinnamon – 4 cm (1½ in) long
4 cloves
4 cardamoms

6 almonds – scald and remove skin)
6 cashew nuts) grind fine
1½ tbsp tomato paste
2 medium sized onions – slice thinly and fry till golden brown
2 tsp *briani* spices (see glossary)
120 g (3 tah) cashew nuts – fry till golden brown and keep aside for garnishing
1 bunch of small coriander leaves for garnishing

Ingredients for rice:

600 g (1 kt) *Basmati* rice
1 small tin evaporated milk
2 tsp yellow food colouring
3 tsp rose water (*ayer mawar*)

Method of preparing chicken:

Combine chicken with garlic-ginger paste, chillies, tomatoes, meat curry powder, big coriander leaves, mint leaves, yoghurt, pepper, salt and ½ cup water from allowance.

Heat *ghee* in a pot then add ingredients for frying and fry till golden brown. Add chicken and fry for a little while, turning it over a few times. Add 1 cup water and bring to the boil.

When boiling, lower heat, add ground almond and cashew nuts, tomato paste, fried onion slices and *briani* spices. Turn over the chicken pieces, then leave to simmer till almost dry. Remove chicken from the pot and set aside. Pour left-over *ghee* into a bowl and leave to cool. This pot will be used for cooking the rice.

Method of preparing rice:

Bring ¾ of a large saucepanful of salted water to boil.

Wash and drain rice.

When water is boiling, put in the rice and boil till it is almost cooked.

When rice is almost cooked, remove from heat and drain off water. Mix evaporated milk with the *ghee* left over from cooking chicken.

Into the pot used to cook chicken, put a layer of rice then a layer of chicken. Repeat till chicken and rice are used up, making sure that rice forms the last layer.

Pour the oil-milk mixture over the rice.

Mix rose water with yellow colouring and sprinkle over rice. Scatter a few mint leaves over rice.

Cook over low heat making sure the pot is well covered till steam emits from pot. Do not uncover pot while rice is cooking.

When serving, garnish rice with fried cashew nuts and small coriander leaves.

RASAM

Ingredients:

½ tbsp black peppercorns)
½ tbsp cumin *(jintan puteh)*) grind coarsely
4 cloves garlic – pound coarsely
1 tsp turmeric powder
1 tsp salt
1 cm (½ in) cube of asafoetida (see glossary)
1 medium tomato
1 sprig curry leaves
1 tbsp tamarind – soak in some of the water from allowance and remove seeds
3½ cups water
1 tbsp oil

2 dried chillies – break into pieces)
1 small sprig curry leaves)
½ onion – finely sliced) to be fried
1 tsp brown mustard seed)

Method:

In a large bowl, combine ground ingredients, garlic, turmeric, salt, asafoetida, curry leaves, tamarind juice and water. Crush tomato in this then mix well.

Heat oil in a deep pot, add ingredients to be fried. When brown add mixture in bowl to this. Cover pot and bring to boil. When steam emits from the pot, soup is ready. This soup may be served hot or cold. Serve with deep fried *papadom.*

SPICY FRIED MUTTON

Ingredients:

450 g (¾ kt) mutton – cut into small cubes
2 green chillies) cut into thin strips lengthwise
2 red chillies)
1 onion – cut into half, then slice thinly
3 tbsp meat curry powder (see pg. 6)
salt to taste
1 sprig curry leaves (*karuvapillai*) – use leaves only
½ cup water
5 tbsp oil
2 cups water

Method:

Combine mutton with all the above ingredients in a bowl (except oil and 2 cups water).
Heat oil, add mutton, and fry till fragrant. Add water, cover and bring to the boil, then continue boiling, stirring now and then.
Lower heat when the amount of liquid is reduced. When curry is almost dry, remove from heat.
Garnish with potato crisps or small coriander leaves.

CARROTS AND GRATED COCONUT

Ingredients:

300 g (½ kt) carrots – peel, slice thinly and cut each slice in half
½ cup grated coconut – grind
1 tsp cumin *(jintan puteh)* powder
2 tsp coriander (*ketumbar*) powder
2 tsp turmeric powder
1 tsp brown mustard seed
1 sprig curry leaves (*karuvapillai*)
salt to taste
¾ cup water
5 tbsp oil

Method:

Combine carrot with coconut, cumin, coriander and turmeric powder in a bowl.
Heat oil, add mustard seeds and curry leaves, and fry till fragrant.
Add carrots, fry till fragrant, then add water and continue cooking till carrots are soft and the consistency slightly wet.

1. *Dalcha*
2. *Nasi Briani*
3. *Pickle*
4. *Spicy Fried Mutton*

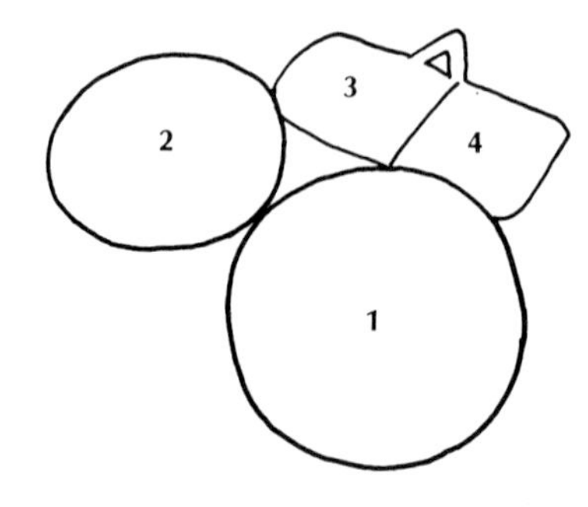

DHAL RASAM

Ingredients:

60 g (2 tah) yellow lentils (*dhal*) – soak for about 3 hours, then wash, remove stones and skin, and drain
6 cups water
1 tsp black peppercorns – grind coarsely
a very small piece of asafoetida about the size of large pea (see glossary)
½ tsp brown mustard seed (optional)
1 sprig curry leaves (*karuvapillai*)
salt to taste

Method:

Put *dhal* and water in a saucepan and boil for about ½ an hour. Drain *dhal* and put aside for use in another recipe. Put water back into saucepan, add pepper, asafoetida, mustard seeds, curry leaves and salt and bring back to the boil.
Continue boiling for another 5 mins.
Serve hot or cold.

Note:

The boiled *dhal* left over may be mixed with sugar and grated coconut and eaten as a sweet.

SAMBAL IKAN BILIS WITH COCONUT

Ingredients:

40 g (1 tah) dried whitebait (*ikan bilis*) – split in half
1 cup grated coconut
5 shallots
3 red chillies
1 piece ginger (size of almond)
1 tsp tamarind – remove seeds
salt to taste
oil for deep frying

Method:

Deep fry *ikan bilis* till crisp, then drain.
Grind coconut, add tamarind pulp and salt and continue grinding till fine. Add chillies, shallots and ginger and when fine, add fried *ikan bilis* and grind coarsely. Put paste in serving dish.

1. *Sambal Ikan Bilis with Coconut*
2. *Hot Kerala Chicken Curry*
3. *Carrots & Grated Coconut*
4. *Dhal Rasam*
5. *Fish Curry with Young Coconut*

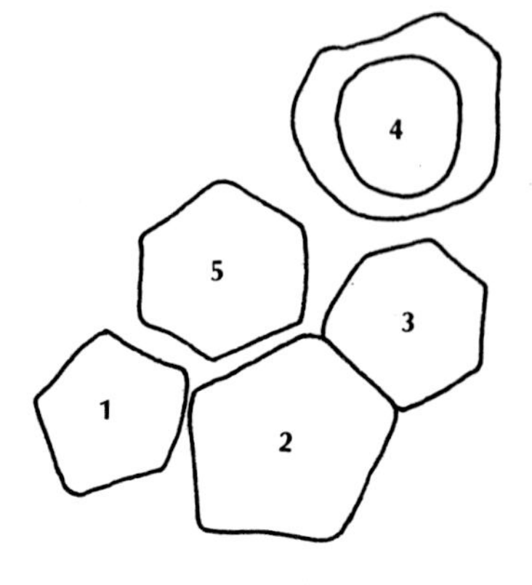

CHICKEN COCONUT MILK CURRY

Ingredients:

1 chicken weighing approximately 1.5 kg (2½ kt) — cut into medium sized pieces
½ medium sized onion — cut into thin wedge shaped slices
1 green chilli) slash
1 red chilli)
salt to taste
4 sprigs big coriander leaves (optional) — cut into pieces
¾ cup water
10 slightly heaped tbsp meat curry powder (see pg. 6)

Grind to a fine paste:

1 piece ginger — size of walnut
4 cloves garlic

In a big bowl, combine chicken with all the above ingredients.

Ingredients for frying:

½ onion — cut into thin slices
1 stick cinnamon — 4 cm (1½ in) long
3 cardamoms
3 cloves

8 tbsp cooking oil
3½ cups water
10 small potatoes — peel and use whole
1 cup grated coconut (optional) — add ½ cup water and extract equal amount of milk

Method:

Heat oil, add ingredients for frying and fry till fragrant. Add chicken and fry till fragrant, then add water and potatoes and bring to the boil. Reduce heat and simmer. When potatoes are half cooked, add coconut milk and bring to the boil. Reduce heat and simmer until potatoes are cooked.

FISH CURRY WITH YOUNG COCONUT

Ingredients:

600 g (1 kt) fish — *ikan tenggiri* or *ikan kurau* (Spanish mackerel, mulloway or any fish steaks) — cut into medium sized pieces
1 young coconut — remove flesh with a spoon and cut into pieces about 2.5 cm (1 in) square

Soak for about an hour:

2 tbsp coriander (*ketumbar*) seeds
1 tsp fennel *(jintan manis)* seeds
½ tsp black pepper corns
½ tsp cumin *(jintan puteh)* seeds
1 piece dried turmeric (size of almond)
10 dried chillies

2 sprigs curry leaves (*karuvapillai*)
1 tsp fish curry spices (*rempah tumis ikan*) (see glossary)
½ tbsp tamarind — mix with 1½ cups water, strain to obtain juice
salt to taste
3 tbsp oil

Method:

Grind soaked spices into a fine paste, adding chillies last. In a bowl, combine fish with curry paste, I sprig curry leaves, salt and coconut flesh.
Heat oil, add 1 sprig curry leaves and fish curry spices, and fry till fragrant. Add fish and coconut flesh, fry till fragrant then add tamarind juice. Cover and bring to the boil over low heat.
When it has boiled, stir, cover again and cook for another 10 minutes.

INDIAN DALCHA

Ingredients:

450 g (¾ kt) mutton ribs — cut into medium sized pieces
1 cup yellow lentils (*dhal*) — soak in water for 2—3 hours

Grind to a smooth paste:

3 slices ginger — each 1 cm (½ in) thick

Method:

Put mutton in a deep pot together with *dhal*, ginger-garlic paste, thickly sliced onion, green and red chillies, turmeric powder, curry leaves, cinnamon, cardamoms, cloves and salt. Add water and boil until meat and *dhal* are cooked.
Add potatoes, bananas, carrots, *brinjal,* mango and

4 cloves garlic
½ big onion – slice thick
2 green chillies) slit half way
1 red chilli)
1½ level tsp turmeric powder
1 sprig curry leaves (*karuvapillai*)
1 stick cinnamon – 3 cm (1¼ in) long
2 cardamoms
3 cloves
salt to taste
8 cups water
300 g (½ kt) small potatoes – peel and cut into half
2 bananas (*pisang kari*) – split in half and cut each half into 3–4 pieces
3 carrots – cut into 4 cm (1½ in) lengths
3 purple eggplants (*brinjals*) – split in half then cut each half into 3–4 pieces
1 green mango – cut into 3 pieces
2 tbsp meat curry powder (see pg. 6) – add 8 tbsp water to make watery mixture
1 cup grated coconut – add ¾ cup water and extract equal amount of thick coconut milk
1½ tbsp tamarind – mix with coconut milk and strain
300 g (½ kt) tomatoes – cut into half
6 tbsp oil
1 bunch small coriander leaves for garnishing

Ingredients for frying:

1 big onion – slice thinly
1 tbsp chilli powder
1 sprig curry leaves (*karuvapillai*)

curry powder mixture; cover pot and boil. Do not uncover until it has boiled for 5 minutes.
Add coconut milk which has been mixed with tamarind and cover pot again. Do not uncover until curry is boiling and the vegetables are cooked.
Add tomatoes and leave for 1 or 2 seconds only, then remove from heat.
Heat oil in frying pan and cook ingredients for frying till golden brown. Pour contents of pan (including oil) into the pot of Dalcha and cover immediately.
When serving, garnish the Dalcha with a few small coriander leaves.

NOTE:
For a hot Dalcha, add more chilli powder when you fry the sliced onions and curry leaves.
If mango is not used, increase amount of tamarind to 2 tablespoons.

SALT FISH CURRY

Ingredients:
150 g (¼ kt) salted *Ikan Kurau* meat (or dried salted cod) – cut into pieces 2 cm (¾ in) thick, soak for a few minutes and wash
230 g (6 tah) salted *Ikan Kurau* bones – cut into 2
10–12 small potatoes – wash skin thoroughly; do not peel
3 purple eggplant (*brinjal*) – cut each in half lengthwise then cut each half into 3 pieces
4 drumsticks – peel and cut into 6 cm (2½ in) lengths
8 long beans – break or cut into 5 cm (2 in) lengths
8 tbsp oil
8 medium sized tomatoes – cut into half

Combine in a large bowl:

10 tbsp fish curry powder (see pg. 6)
2 tbsp grated coconut – grind into a fine paste
2 green chillies) break into 3 pieces
1 red chilli)
1 sprig curry leaves *(karuvapillai)*
½ onion – slice thickly
2 tbsp tamarind – mix with 3 cups water and strain to get tamarind juice
2 cloves garlic – slice thickly
salt to taste
1 whole tomato – crush

Ingredients for frying:

½ onion – slice finely
1 tbsp fish curry spices (*rempah tumis ikan*) (see glossary)
1 small sprig curry leaves (*karuvapillai*)

Method:
Heat oil, add ingredients for frying. When fragrant, add potatoes and fry for 5 minutes. Add ingredients which have been combined, and *brinjal,* long beans and drumsticks. Cover, bring to the boil and continue boiling till potatoes are nearly cooked. Add salt fish and bones, reduce heat and cook for another 10 minutes. Add tomato halves, leave for a minute then remove from heat.

Note:
If mangoes are available, add 2 green mangoes, cut into small pieces, together with the other vegetables. Reduce amount of tamarind if mango is used. No salt is needed for this curry as the salt fish will give it sufficient taste.

Note:
Serve with deep fried *chilli tairu* available from any Indian spice shop.

Noodles and Singapore Dishes

MEE SOTO

Ingredients:
1 chicken weighing about 1.75 kg (3 kt)
6 cups water
salt to taste
600 g (1 kt) fresh yellow noodles
300 g (½ kt) bean sprouts – remove tails

Grind into a fine paste:
1½ tbsp coriander (*ketumbar*) seeds
1 tsp cumin *(jintan puteh)* seeds
1 tsp fennel *(jintan manis)* seeds
1 stalk lemon grass (*serai*) – slice
1 thick slice *lengkuas*
4 candlenuts (*buah keras*)
8–10 shallots
1 tsp white peppercorns

For garnishing:
10 shallots – cut into fine slices and fry till golden brown
spring onions and celery – chop fine
potato cutlets – fry
shredded chicken

Chilli Padi Sauce:
¼ cup birds-eye chilli (*chilli padi*) – wash and remove stalks; grind fine and pour some dark soya sauce over it.

Method:
Put whole chicken, ground paste, water and salt in a large pot. Bring to the boil. When boiling, remove chicken and keep stock aside to use as soup for the Mee Soto.
Put the chicken on a tray and roast for about 20–30 minutes at 190°C (375°F) until a light brown.
Shred chicken when cool.
Bring a large pan of water to the boil. When water is actively boiling, put in noodles and bring back to the boil. Continue boiling till noodles are cooked, then remove noodles.
In the same water, scald the bean sprouts.

To serve:
Put the noodles in a dish, sprinkle bean sprouts and shredded chicken over it then add soup and garnishing. Serve with *Chilli Padi Sauce.*

1. *Fried Kway Teow*
2. *Fried Beehoon Indian Style*
3. *Mee Soto*
4. *Mee Siam Lemak*
5. *Baked Macaroni*

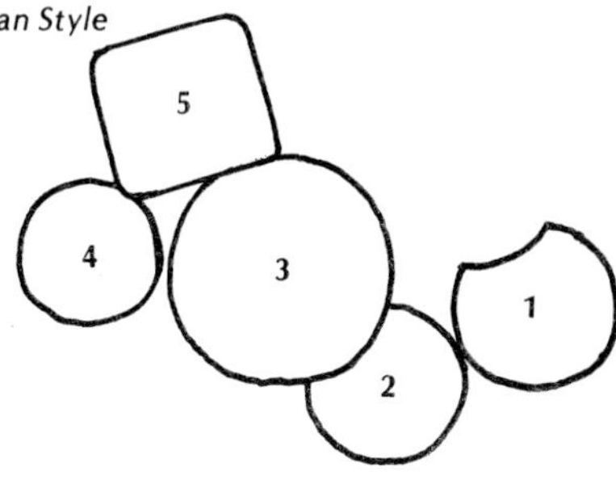

FRIED BEE HOON — INDIAN STYLE (Rice Noodles)

Ingredients:

Grind to a fine paste:

- 1 onion
- 1 clove garlic
- 16 dried chillies – break each into 3–4 pieces and soak

4 tbsp oil
2 hard beancurd (*taukwa*) – cut into small squares
salt to taste
2 tbsp tomato sauce
½ tbsp dark soya sauce
4 medium sized tomatoes – slice thickly
2 red chillies) slice
2 green chillies)
2–3 eggs – beat lightly
300 g (½ kt) spring greens (*chye sim*) – cut into 3 cm (1¼ in) lengths
300 g (½ kt) bean sprouts – remove tails
3 medium sized potatoes – boil and cut into slices 0.5 cm (¼ in) thick
230 g (6 tah) rice vermicelli (*bee hoon*) – soak for 15–20 mins then cut into shorter strands

Method:

Heat oil, add ground ingredients and fry till fragrant. Add *taukwa,* salt, tomato sauce and dark soya sauce and stir. Add tomatoes and sliced chillies stir, then push ingredients to the side of the pan.

Pour the eggs into the middle and let eggs cook. Mix with other ingredients in the pan, then add *chye sim,* bean sprouts and boiled potatoes, stirring after each addition. Finally, add *bee hoon,* stir, mixing it with all the other ingredients.

Serve with sliced cucumber, tomato sauce and cut lime.

MEE SIAM LEMAK

Ingredients:

80 g (2 tah) salted *ikan kurau* (or dried salt cod)
4 pieces hard beancurd (*taukwa*) – cut each into small cubes and rub with salt
25 dried chillies – cut and soak
10 shallots
1 clove garlic
2 tbsp salted soyabeans (*taucheo*)
300 g (½ kt) thick dried white noodles (*beehoon kasar*) – soak for about 1–2 hours then cut into shorter lengths
300 g (½ kt) shrimps – remove shell
cooked coconut crumbs made from milk of ½ coconut (see pg. 107)
300 g (½ kt) bean sprouts – remove tails
salt to taste
12 tbsp oil

For garnishing:

- 4 hard boiled eggs – sliced
- a bunch of coarse chives (*kuchai*) – cut into 3 cm (1¼ in) lengths
- a few small limes – cut into half

Method:

Roast the pieces of *ikan kurau* in a dry, hot pan until fish is slightly brown. Remove from heat, flake it then grind into a fine paste with chillies, shallots, garlic and *taucheo.*

Heat oil, fry *taukwa* cubes till slightly brown, then put aside in a dish. In the same oil, fry the ground ingredients, adding salt, till fragrant. Add shrimps, stir fry for a few minutes then add coconut crumbs. Stir, then add noodles and stir fry, mixing the noodles with the other ingredients. Add bean sprouts, mix and stir fry for a few more minutes then remove from heat.

Garnish with fried *taukwa,* sliced hard boiled eggs, *kuchai* and limes.

BAKED MACARONI

Ingredients:

300 g (½ kt) minced beef
½ tsp pepper
1 tsp chilli powder
½ tsp turmeric powder
2 tsp coriander (*ketumbar*) powder

Method:

Mix minced beef with ginger-garlic paste, pepper, chilli, turmeric, coriander powder and salt.

Heat oil, add beef, fry until liquid is absorbed, then remove from heat.

Add sliced chilli to cooked beef.

salt to taste
2 red chillies – thinly sliced, slantwise
1 slice ginger (2 cm or ¾ in thick)) grind to a fine
1 clove garlic) paste
2 tbsp cooking oil
230 g pipe macaroni
Chop fine:
  2 spring onions
  3 sprigs small coriander leaves
  3 sprigs big coriander leaves
½ tbsp margarine
3 eggs – beaten with a little salt

Bring a pan of water to the boil. When actively boiling, add macaroni and cook for about 12 minutes.
Spread the margarine in an ovenproof dish, then put about two-thirds of the macaroni to form the first layer. Spread meat, chopped spring onions and coriander leaves over macaroni to form a second layer; spoon one beaten egg over this. Finally, cover this layer with the rest of the macaroni and spoon two beaten eggs over this. Dot the top layer with margarine and bake in moderate oven for about 20 minutes.
Serve with cucumber salad.

FRIED KWAY TEOW

Ingredients:
600 g (1 kt) fresh rice flour noodles (*kway teow*)
120 g (3 tah) beef – thinly sliced
300 g (½ kt) shrimps – peel
20 fish balls – cut in half
300 g (½ kt) spring greens (*chye sim*) – cut into 5 cm (2 in) lengths
Grind coarsely:
  8 shallots
  2 small cloves garlic
Mix together:
  2 tsp cornflour
  3 tbsp light soya sauce
  1 tsp pepper
  1 cup water
6 tbsp oil
salt to taste

Method:
Heat oil, add ground ingredients and fry till fragrant. Add beef, shrimps, fish balls, *chye sim* and salt, stir frying after each addition. Add cornflour mixture. When bubbling, add *kway teow* and stir, mixing all ingredients well. Serve immediately.

CHICKEN BEE HOON BRIANI SYTLE

Ingredients:
450 g rice vermicelli *(bee hoon)* – soak in water till soft
2 tsp yellow food colouring – sprinkle over *bee hoon* and mix evenly
2 tsp white peppercorns – grind fine
1 chicken weighing about 1.5 kg (2½ kt) – cut into pieces and season with ground pepper; leave for 10-15 mins and deep fry till golden brown
3 tbsp *ghee* or cooking oil
Grind to a smooth paste:
  2 slices ginger – each 1 cm (½in) thick
  3 cloves garlic
  6 dried chillies – soak
salt to taste
4 tomatoes – half and slice thinly
1 red chilli) slit into two
1 green chilli)
1 tbsp meat curry powder (see pg. 6)
1 bunch big coriander leaves
230 ml (l scant cup) water
For garnishing:
  10 shallots – slice thinly and deep fry till golden brown
  mint leaves
  small can of green peas – drain

Method:
Heat *ghee* or oil in a frying pan and fry ground ingredients until fragrant. Add salt, tomatoes, chillies, curry powder and coriander leaves. Fry for 2–3 minutes then add water.
When boiling, add fried chicken and mix thoroughly, making sure chicken is well coated with fried ingredients. Remove chicken, leaving the spices in the pan. Add coloured *bee hoon* to pan, stir fry and mix well, adding salt if necessary. Put chicken back in pan and mix.
Serve on a bed of lettuce and garnish with green peas, fried shallots and mint leaves.
Serve with pineapple and cucumber salad if desired.

CRISPY MUTTON CHOPS

Ingredients:
6 lamb or mutton chops
whites of 3 eggs
1 tbsp honey
1 tbsp dark soya sauce
1½ tbsp fine bread crumbs
oil for deep frying

Method:
Boil mutton with 1 tsp salt and 1 tsp pepper for 30 minutes in just enough water to cover.
Combine egg white with honey, soya sauce and bread crumbs. Dip the boiled chops into mixture and deep fry.
Serve with lemon wedges or mint sauce, potato chips and green peas.

GULAI NANAS

Ingredients:
2 tbsp oil
Grind to a fine paste:
- 10 dried chillies – soak
- 1 piece dried shrimp paste *(blacan)* – 2 x 2 x 0.5 cm (¾x ¾x ¼ in)
- 8 shallots
- 1 piece fresh turmeric – about 2 cm (¾ in) long

1 stalk lemon grass *(serai)*) bruise
1 small piece *lengkuas*)
1 tsp salt
1 tbsp sugar
½ almost-ripe pineapple – cut into two pieces lengthwise and then cut each piece into wedges about 1 cm (½ in) thick
2 cups water
1 tsp tamarind – mix with ½ teacup water and extract juice
5 pieces *ikan tenggiri* (Spanish mackerel or bonito)

Method:
Heat oil in a pot and fry the ground ingredients, *serai* and *lengkuas* till fragrant. Add salt, sugar, pineapple and water. Cover pot and boil for 3-5 minutes. Add tamarind juice and *ikan tenggiri*, cover and boil a few minutes over low heat.
As soon as fish is cooked, remove from heat.

1. *Opor Sotong*
2. *Fried Long Beans with Grated Coconut*
3. *Savoury Rice with Curried Lamb*
4. *Chicken Beehoon Briani Style*
5. *Gulai Nanas*

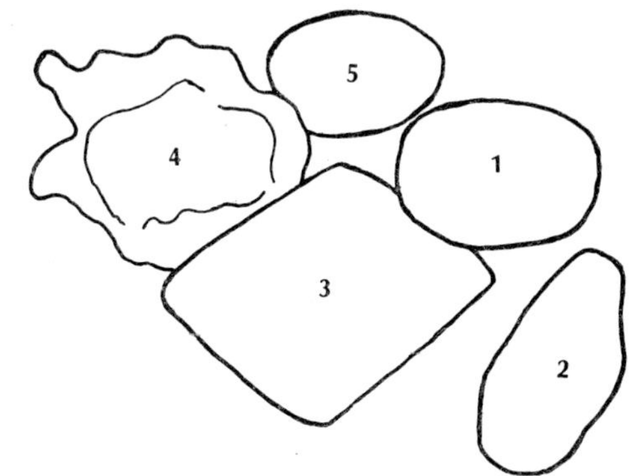

SAVOURY RICE WITH CURRIED LAMB

Ingredients for Curried Lamb:
450 g (¾ kt) lamb – cut into small squares
2 tbsp meat curry powder (see pg. 6)
2 tsp tomato paste
1 tsp peppercorns – grind coarsely
1 large onion) cut into small squares
2 tomatoes)
1 tbsp plain yoghurt
salt to taste
Grind to a fine paste:
1 slice ginger – 1 cm (½ in) thick
2 cloves garlic
1 red chilli
1 green chilli
Combine lamb with all the above ingredients and marinate for a while.
2 tbsp *ghee*
½ onion – slice thinly
2 cups water
4 potatoes – peel and cut into 6 pieces lengthwise
For garnishing:
20-30 cashew nuts – fry in oil till golden brown
1 small bunch big coriander leaves
1 small bunch mint leaves

Method:
Heat *ghee* in pot and fry sliced onion till golden brown; add marinated meat and stir-fry for 1 or 2 minutes until fragrant. Add water, bring to the boil then add potatoes. Cover pot and simmer till meat is tender and the gravy fairly thick.

Ingredients for rice:
600 g (1 kt) *Basmati* rice – wash and drain
1.6 litres (6½ cups) water
1 tsp salt
1 tbsp *ghee*
For garnishing:
4 tbsp raisins) fry in *ghee*
sliced almonds)

Method:
Bring water to the boil, add salt and rice and boil for 10-15 minutes checking often to see if rice is cooked. When rice is cooked, drain, then rince rice in the strainer by pouring more boiling water over it to remove any excess starch. Mix *ghee* with rice. Make a ring with rice in an ovenproof dish. Put curried lamb in hollow of ring and bake in a moderate oven for 10 minutes. Garnish rice with raisins and almonds, and curry with cashew nuts, coriander and mint leaves.

EGG AND LAMB (OR BEEF) KEBAB

Ingredients:
230 g (6 tah) minced lamb or beef
3–4 sprays big coriander leaves – chop fine
1 tsp pepper
1 tsp chilli powder
1 tsp coriander (*ketumbar*) powder
¼ tsp turmeric powder
1 slice toast – crush into fine crumbs
2 egg yolks
2 green chillies – chop into small pieces
1 chopped onion (optional)
5 hard boiled eggs
oil for deep frying

Method:
Combine all ingredients (except hard boiled eggs) in a bowl and mix well.
Divide mixture into five portions. Wrap one portion around each egg, then deep fry till golden brown. Cut eggs in half, serve on a bed of lettuce with raw onion rings, and mint or chilli sauce.

FRIED LONG BEANS WITH GRATED COCONUT

Ingredients:
2 tbsp oil
½ onion – slice thinly
1 green chilli – slice diagonally
1 tsp turmeric powder
1 small sprig curry leaves (*karuvapillai*)
½ tsp brown mustard seeds
½ cup grated coconut – grind coarsely
salt to taste
½ teacup water
300 g (½ kt) long beans – cut into 2 cm (¾in) lengths

Method:
Heat oil in frying pan and fry onion, chilli, turmeric powder, curry leaves and mustard seeds till light brown. Add ground coconut, salt and keep frying till fragrant. Add water; stir, then add long beans and cook till beans are tender stirring now and then while cooking.

OPOR SOTONG

Ingredients:
4 tbsp oil
Grind to a smooth paste:
1 clove garlic
1 medium sized onion
1 piece fresh turmeric – about 0.5 cm (¼ in) long
12 dried chillies – soak
2 tbsp coriander *(ketumbar)* powder) add to ground
1 tsp cumin *(jintan puteh)* powder) ingredients
1 stalk lemon grass *(serai)*)
1 piece *lengkuas* – 1 cm (½ in) thick) bruise
450 g (¾ kt) grated coconut (put aside 2 tbsp for roasting) – add 2 cups water and extract 3 cups milk
salt to taste
2 fragrant lime leaves *(daun limau perut)*
600 g (1 kt) medium or big cuttle fish *(sotong)* – separate head from body, remove bag of fluid, wash and drain

Method:
Heat oil in pot and fry ground ingredients together with *serai, lengkuas,* roasted coconut, salt and lime leaves till fragrant. Add coconut milk. When boiling, stir, reduce heat and simmer until gravy becomes thick, stirring occasionally.
Add *sotong* and cook for 5 minutes. Do not overcook as *sotong* will become rubbery.

HONEYDEW FRIED CHICKEN

Ingredients:
1 chicken weighing about 1.5 kg (2½ kt) – cut into pieces of average size
1 tsp pepper
2 tbsp honey
2 tbsp dark soya sauce
oil for deep frying
120 g (3 tah) cashew nuts

Method:
Combine chicken with pepper, honey and soya sauce and leave to marinate for 1 hour. Drain chicken keeping marinade for soup. Heat oil and deep fry chicken till golden brown.
Deep fry cashew nuts till golden brown. When serving, sprinkle cashew nuts over chicken and garnish with slices of cucumber. Serve hot.

SOUP

Ingredients:
5 shallots)
1 clove garlic) grind coarsely
1 cm (½ in) ginger)
2 chicken legs
gizzard and liver from 1 chicken
chicken neck and bones
3½ cups water
salt to taste
½ tsp pepper
small lettuce leaves

Method:
Combine all the above ingredients in a pot, and add the left-over marinade from Honeydew Fried Chicken. Bring to the boil then leave to simmer for a little longer. Drop in a few small lettuce leaves before serving.

GINGER PICKLE

Ingredients:
150 g (¼ kt) young ginger – peel and slice paper thin
3 tbsp granulated sugar
½ cup local vinegar

Method:
Either mix all ingredients and leave to stand for at least one day before using, **or** boil vinegar and sugar, cool then add ginger slices.
Store in a glass bottle.
Serve with preserved quails' eggs.

FRIED POH PIAH
(Spring Rolls)

Ingredients:

Grind coarsely:

4 tbsp salted soya beans (*taucheo*)
2 cloves garlic

4 tbsp oil
½ cup water
300 g (½ kt) small shrimps – remove shell
230 g (6 tah) crabmeat
300 g (½ kt) bamboo shoot – shred (canned bamboo shoots may be used)
150 g yam bean (*bangkwang*) – shred
300 g (½ kt) bean sprouts – remove tails
4 pieces hard beancurd (*taukwa*) – cut into small squares and deep fry
a few stalks big coriander leaves – cut into 1 cm (½ in) lengths
600 g (1 kt) fresh *poh piah* skin

Method:

Heat oil, add ground *taucheo*-garlic paste and stir fry for a minute then add water and continue to fry till fragrant.
Add shrimps and crabmeat, fry for a few minutes, then add bamboo shoot, *bangkwang* and bean sprouts, stirring after each addition for a few minutes. Remove from heat and sprinkle fried *taukwa* over it. Leave to cool.

To wrap:

Put about a tablespoon of cooked filling on each skin, add one or two pieces of coriander leaves and wrap to form a roll. Seal the edge with a little water.
Heat sufficient oil for deep frying and fry rolls till golden brown.
Serve, garnished with tomato and cucumber slices, with chilli sauce.

Note:

If *poh piah* skins are not available, spring roll skins (which are coarser) can be substituted. Seal ends with flour and water paste.

1. *Egg & Lamb Kebab* **2.** *Crispy Mutton Chops.* **3.** *Chocolate Sponge Pudding* **4.** *Indian Mutton Soup*

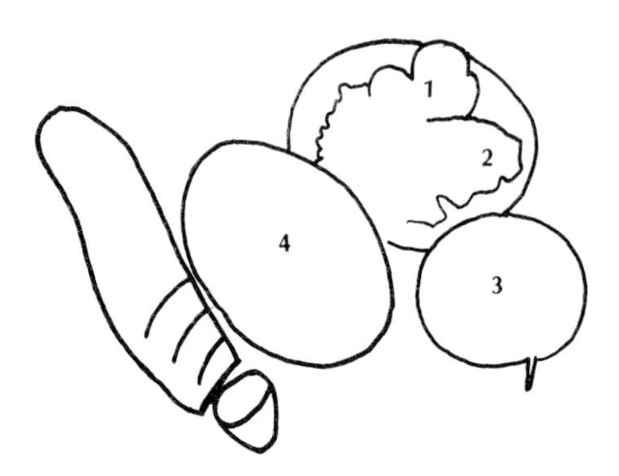

STUFFED BEEF ROLL

Ingredients:

2 large carrots – peel and grate coarsely
4 large potatoes – boil and peel (add a pinch of salt to water for boiling)
¼ tsp salt
½ tsp pepper
8 large thin slices of topside beef
4 tsp butter or margarine

Method:

Mash potatoes, add pepper and salt.
Lay a piece of beef flat on a board, spread a thin layer of mashed potatoes over it. Put 1 tbsp grated carrot over the potato and fold to form a roll. Secure with thread. Repeat till all the pieces of beef are used.
Spread 2 teaspoons of butter or margarine on a baking tray. Place rolls in tray and dot each roll with a little butter or margarine. Put the tray under the grill. After about 10 minutes turn rolls over, grill for another 5–8 minutes till meat is brown.
Remove beef rolls from tray and keep tray aside.

Note:

Beef rolls may be cooked for about 20 minutes in a moderately hot oven (200°C or 400°F).

SAUCE FOR BEEF ROLL

Ingredients:

4 tbsp tomato sauce
2 cups water
½ tsp pepper
salt to taste
1 tsp golden bread crumbs
120 g (3 tah) green peas
4–5 medium sized tomatoes – quartered

Method:

Add ½ cup water to baking tray. Stir to mix with pan juices, then pour into a saucepan and add tomato sauce, rest of water, bread crumbs, pepper and salt. Bring to the boil, then add green peas and tomato quarters. Bring back to the boil and remove from heat. Pour sauce over rolls just before serving.

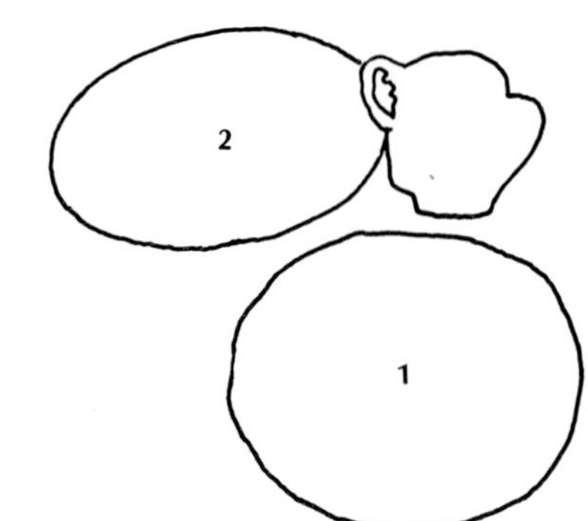

1. *Stuffed Beef Roll*
2. *Orange Jelly*

MAGIC JELLY

Ingredients:
40 g (1 tah) *agar-agar* strands
450 g (¾ kt) granulated sugar
200 ml fresh milk
2 fragrant screwpine leaves (*daun pandan*) — wash and tie into a knot
6 cups water
1 tsp rose essence
yellow, green, red and blue food colouring

Method:
Soak *agar-agar* for 1 hour then drain.
Put *agar-agar, pandan* leaves and water in a large pot. Bring to the boil, then continue boiling till *agar-agar* dissolves. Add sugar, stir and when sugar dissolves, strain jelly into a large bowl.
Add milk and essence to jelly and stir.
Put 2 large ladles of jelly in each of four small shallow bowls, leaving rest in large bowl. Add a different colour to each small bowl and chill coloured jelly in fridge. Leave the white mixture at room temperature to prevent it setting before coloured jellies.
When the bowls of coloured jelly have set, cut them into small cubes, add to the white mixture and stir to mix. Pour into jelly moulds and chill.

ORANGE JELLY

Ingredients:
1 packet (40 g) *agar-agar* strands — soak for ½ an hour
10 cups water
1 can evaporated milk (410 g/369 ml)
2 eggs
450 g (¾ kt) granulated sugar
2 tsp orange essence
1 tsp yellow food colouring

Method:
Boil *agar-agar* in water. When dissolved, strain into another saucepan. Mix the milk with eggs, half the amount of sugar and essence in a large bowl. Dissolve the rest of the sugar in a large saucepan and cook till sugar becomes brown and cystallizes.
Combine milk mixture with jelly and add to crystallized sugar. Cook, stirring all the time, until it boils, then continue cooking for a minute or two. Remove from heat and pour into large jelly moulds or trays. Leave to cool, then chill in fridge.

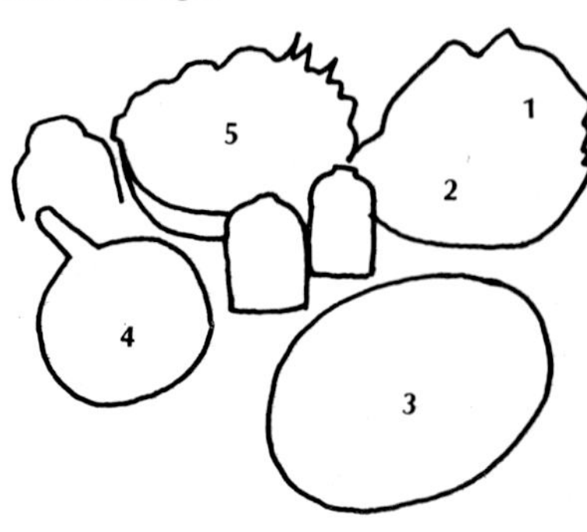

1. *Fried Poh Piah*
2. *Ginger Pickle*
3. *Dry Chap Chye*
4. *Soup*
5. *Honeydew Fried Chicken*

INDIAN MUTTON SOUP

Ingredients:
450 g (¾ kt) mutton – cut into small squares
2 tsp white pepper
4 tbsp coriander (*ketumbar*) powder
1 fresh green chilli – slash
salt to taste
Grind to a smooth paste:
- 2 slices ginger – each 1 cm (½ in) thick
- 3–4 cloves garlic
- 2 tsp white poppy seeds (*kas kas*)
- 1 tsp cumin *(jintan puteh)*

4–5 tbsp oil
Ingredients for frying:
- 1 stick cinnamon
- 3–4 cloves
- 2–3 cardamoms
- 1 medium sized onion – slice thinly

7–8 cups water
For garnishing:
- 4–5 spring onions – chop coarsely
- 8–10 shallots – peel, slice thinly and fry till golden brown; set aside

Method:
Combine mutton with ground ingredients, pepper, coriander, green chilli and salt in a bowl and mix evenly.
Heat oil, then add ingredients for frying. Fry till golden brown, add meat and fry till fragrant then leave to cook for 5–8 minutes. Add water and boil till meat is tender.
Garnish with fried shallots and chopped spring onion. Serve with French loaf or toast.

PINEAPPLE PUTCHREE

Ingredients:
1 medium sized almost-ripe pineapple – peel, quarter, remove hard core, then cut into wedges about 1 cm (½ in) thick
3 tsp turmeric powder
3 red chillies) slit lengthwise
3 green chillies)
5 tbsp sugar
Ingredients for frying:
- 1 stick cinnamon 6 cm (2½ in)
- 4 cardamoms
- 6 cloves
- 1 onion – slice thinly
- 1 piece ginger (size of walnut) – slice thinly and cut into matchstick lengths

4 tbsp oil
salt

Method:
Bring pineapple to the boil with turmeric powder, then cook for 10 minutes. Remove from heat and drain. Discard water.
Heat oil, add ingredients to be fried. Fry till fragrant, add pineapple, sugar, chillies and a pinch of salt stirring after each addition. Continue to cook for another 5 minutes.

MYSORE FRIED CHICKEN

Ingredients:
1½ kg (2½ kt) chicken – remove skin and cut into small pieces
2 onions – cut into half and slice thinly
2 green chillies) cut into thin strips
1 red chilli)
2 tbsp coriander (*ketumbar*) powder
1 tbsp chilli powder
1 tsp turmeric powder
salt to taste
10 stalks big coriander leaves – chop
2 cups water
8 tbsp oil

Method:
Combine chicken with all the above ingredients (except oil) and 1 cup water from allowance in a big bowl.
Heat oil, add chicken, fry till fragrant. Add rest of water, cover and leave to boil over a low heat until chicken is dry and brown, stirring occasionally.
Garnish with small coriander leaves and potato crisps.

DRY CHAP CHYE

Ingredients:

300 g (½ kt) shrimps – peel
2 hard beancurd (*taukwa*) – deep fry lightly and cut each into 9 squares
60 g (1½ tah) beancurd stick (*foo chok*) – cut into 5–6 cm (2–2½ in) lengths and soak to soften
40 g (1 tah) dried lily buds (*kim chiam*) – soak and remove tiny hard portion at one end of stalk
20 g (½ tah) dried mushrooms – soak in hot water then cut off stems and cut each mushroom in half
40 g (1 tah) dried beancurd sheets (*tim chok*) – soak and cut into fine strips
20 g black fungus (*telinga tikus*) – soak
2 small bundles bean thread vermicelli (*sohoon*) – soak
300 g (½ kt) cabbage – cut into pieces
½ cup oil

Grind coarsely:

3 tbsp salted soya beans (*taucheo*)
2 cloves garlic
10 shallots

6 tbsp oil
2 tbsp light soya sauce

Method:

Heat about ½ cup of oil and fry each of the above items, separately except for *sohoon* and shrimps, and keep aside. They are to be only lightly browned.
Heat 6 tbsp oil, fry ground ingredients till fragrant.
Add shrimps, fry, add soya sauce and all the fried ingredients except cabbage and stir fry. Add cabbage, then *sohoon*, stirring after each addition, and stir fry for a few more minutes before removing from heat.

TOMATO RICE

Ingredients:

600 g (1 kt) *Basmati* rice – wash and drain
water – equal in volume to rice used

Grind to a smooth paste:

3 cloves garlic
2 slices ginger (each 2 cm or ¾ in thick)

Ingredients for frying:

½ onion – thinly sliced
1 length cinnamon 5 cm (2 in)
4 cloves
3 cardamoms

1 fragrant screwpine leaf (*daun pandan*)
1 tin tomato juice (approx. 1 cup)
1 tbsp plain yoghurt
2 tbsp evaporated milk
4 tbsp *ghee*
salt to taste

Method:

Mix tomato juice and yoghurt with water and milk.
Heat *ghee*, add ingredients for frying, garlic-ginger paste and *pandan* leaf. When onions are brown, add tomato-water-milk mixture and salt. Cover pot and bring to the boil.
When water is absorbed, lower heat and continue cooking, making quite sure the pot is well covered.
When steam emits from pot, the rice is cooked.
When serving, garnish rice with green peas and shredded omelet.

Note:

The cooking takes about 45 mins.
This amount of rice will serve 6–8 people.

CHOCOLATE SPONGE PUDDING

Ingredients:

290 g (½ kt or 10 oz) butter or margarine
290 g (½ kt or 10 oz) caster sugar
6 eggs
290 g (½ kt or 10 oz) plain flour sieved with 2 tsp baking powder
2 tbsp cocoa
3–4 tbsp warm water

Method:

Cream butter and sugar until light and fluffy. Beat in eggs, one at a time.
Fold in the sieved flour and cocoa, adding sufficient water to make a soft dropping consistency so that mixture drops readily from spoon.
Pour mixture into a greased pudding basin, filling it only three-quarters full to allow room for pudding to rise.
Steam for 1–1¼ hrs.
Turn out on a dish. If it appears to stick, loosen edges with a blunt knife. Serve at once with custard sauce.

Custard Sauce

Ingredients:
1 egg
1 tbsp sugar
250 ml (1 cup) milk
2 tsp custard powder

Method:
Beat egg and sugar lightly in a saucepan.
Pour in milk, add custard powder, mix well then cook over low heat, stirring all the time, till it boils and thickens.

BEEF SAMAK

Ingredients:
300 g (½ kt)topside beef or round steak — cut into two pieces
2 medium sized tomatoes — slice thinly
Grind to a fine paste:
 1 slice ginger 2.5 cm (1 in) thick
 4 cloves garlic
 8 dried chillies — soak to soften
1 tsp coriander (*ketumbar*) powder
1 tsp cumin *(jintan puteh)* powder
½ tsp white pepper powder
1 tbsp plain yoghurt
1 cup water
2 tsp tomato puree — mix with a little water
6 tbsp oil
salt to taste
Ingredients for frying:
 1 small onion — slice thinly
 1 stick cinnamon 3 cm (1¼ in) length
 3 cloves
 3 cardamoms
 2 tbsp cooked peas

Method:
In a bowl, combine beef with sliced tomatoes, ground ingredients, coriander, cumin and pepper powder, yoghurt and salt.
Heat oil, add ingredients for frying. Fry till onions are brown then add beef and fry till fragrant. Add water, cover and boil. When meat is tender, add puree, stir and cook for another 10 minutes. Gravy should be of a thick consistency.
Garnish with cooked green peas.

1. *Mysore Fried Chicken* **2.** *Pineapple Putchree* **3.** *Tomato Rice* **4.** *Beef Samak*

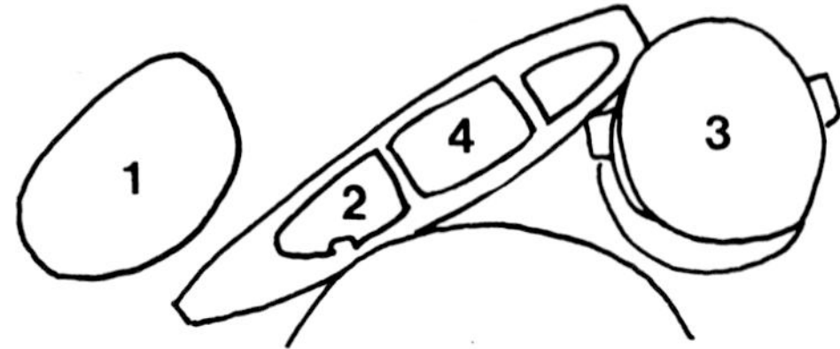

BAMIAH (ARAB DISH)

Ingredients:

600 g (1 kt) topside beef – cut into large cubes

Roast in dry pan over low heat:

3 tbsp coriander (*ketumbar*)
2 tsp cumin *(jintan puteh)* seeds
1 tsp fennel *(jintan manis)* seeds
12 dried chillies

1 tsp white peppercorns
1 piece dried turmeric (size of almond)

Grind to a fine paste:

4–5 cloves garlic
1 piece ginger (size of walnut)

salt to taste
3 tbsp *ghee* or 5 tbsp oil
2 cups water
1 tbsp tomato paste

Ingredients to be fried:

1 medium sized onion – slice thinly
1 stick cinnamon 5 cm (2 in)
3 cardamoms
3 cloves

4 tbsp *ghee*
300 g (½ kt) small ladies fingers (okra) – cut off top and tail and slash half way from tail end up
300 g (½ kt) small tomatoes – cut off top
green peas for garnishing

Method:

Grind coriander, cumin, fennel, dried chillies and turmeric finely, adding peppercorns last.
Combine meat in a bowl with ground ingredients, ginger-garlic paste and salt.
Heat *ghee* or oil, add ingredients to be fried. When onions are a little brown, add meat and fry until fragrant, then add water combined with tomato paste. Cover and simmer till meat is tender, stirring occasion -ally.
Heat 4 tbsp *ghee,* add ladies fingers and fry for about 8 minutes. When meat is tender and gravy of a thick consistency add ladies fingers together with tomatoes. Stir and remove from heat. Garnish with green peas.

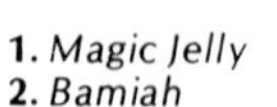

1. *Magic Jelly*
2. *Bamiah*

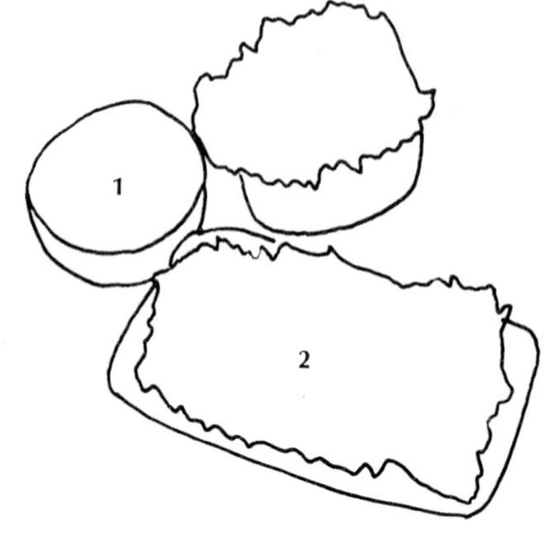

Cocktail Savouries

CHICKEN LEMPER

Ingredients:

300 g (½ kt) glutinous rice (*beras pulot*) – wash and soak in water, to which 1 tsp salt has been added, for 2 hours
½ cup water
1 breast or ¼ of a chicken – rub with salt and steam (Keep stock after steaming)

Chop fine:

1 green chilli
1 red chilli
1 stalk lemon grass (*serai*)
4 shallots

1 cup grated coconut – add ½ cup water and extract ½ cup milk
1 tsp chilli powder
2 heaped tsp coriander (*ketumbar*) powder
½ level tsp turmeric powder
1 tsp sugar
salt to taste
2 tbsp oil

Method:

Drain rice, put in a tray with ½ cup water and steam. When cooked, take it out, stir and return to steamer for another 10 minutes.
Shred cooked chicken and chop into small pieces. Combine shredded chicken with chopped chillies, *serai,* shallots, chilli powder, coriander powder, turmeric powder, salt, sugar and chicken stock. Heat oil, add chicken mixture and fry till fragrant. Add coconut milk and cook till dry stirring continuously during the final stage of cooking. Put half the cooked glutinous rice in a shallow tray, press with the back of a spoon to make the layer compact and level, then spread the cooked chicken over this layer. Cover chicken with the rest of the rice. Press again with the spoon to level the rice and make it compact. Leave to cool then cut into pieces.

1. *Fried Ikan Bilis*
2. *Murukku*
3. *Samosa*
4. *Chicken Kebab*
5. *Vegetable Puffs*
6. *Minced Meat Tarts.*
7. *Satay Flavoured Cockles & Fish Balls*
8. *Spicy Cuttlefish*

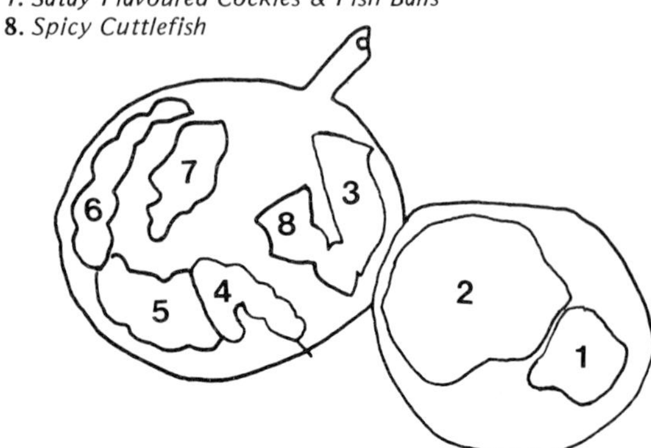

MURUKKU

Ingredients:

600 g gram flour *(besan)*
1 tbsp omum seeds — sieve and remove stones
1 slightly heaped tbsp chilli powder
2 slightly heaped tsp turmeric powder
1 tsp salt, mix with 300 ml water
oil for deep frying

Method:

Combine all the above dry ingredients in a large bowl, add salt water and mix till dough is smooth.
Put a handful of dough into murukku container.
Heat oil and lower murukku strands into hot oil by turning handle of the container. Fry till crisp.
Repeat till all the dough is used up.
Store in air tight container.

MINCED MEAT TART

Ingredients for filling:

120 g (3 tah) minced beef, lamb or pork
4 cloves garlic) grind to a fine paste
1 thick slice ginger 2 cm (¾in))
½ tsp black pepper
3 heaped tsp meat curry powder (see pg. 6)
salt to taste
6 sprigs big coriander leaves —chop
2 tbsp oil

Ingredients for pastry:

230 g (6 tah) plain flour
180 g (4½ tah) margarine
2 egg yolks
1-2 drops yellow food colouring — (optional)

Method for filling:

Combine minced meat with ginger-garlic paste, pepper, curry powder and salt.
Heat oil, add meat, cook till almost dry then add coriander leaves.
Stir and remove from heat, Leave to cool.

Method for pastry:

Sieve flour into a bowl. Rub margarine into flour, add egg yolks and colouring and bind to form a dough.
Take half the dough, roll out pastry to 0.5 cm (¼ in) thick on floured board.
Cut into rounds with a tart cutter. Fill centre with a teaspoonful of cooked meat. Repeat with the rest of the dough.
Bake in a moderately hot oven for 20 minutes.

CHICKEN KEBAB

Ingredients:

600 g (1 kt) chicken — separate meat from bones and cut meat into small cubes

Grind to a fine paste:

1 tsp cumin *(jintan puteh)* seeds) soak for ½ hour
1 tsp black peppercorns)
2 slices ginger — each 1 cm (½ in) thick
3 cloves garlic
2 red chillies
coriander (*ketumbar*) powder
taste
lain yoghurt or juice of ½ lemon
il

Method:

Combine ground ingredients, coriander powder, salt, and yoghurt or lemon juice with chicken. Leave to marinate for about 10 minutes.
Heat oil, then cook till chicken is tender and fairly dry.
Alternatively, put marinated chicken on a tray brush with oil and grill.

To serve:

Put chicken pieces with pickled onions through cocktail sticks or little skewers.

SAVOURY FRIED TAPIOCA CHIPS

Ir
6(
2 tapioca — peel and cut into chips
1 t(*ketumbar*) seeds) soak and grind
1 t *puteh)* seeds)
salt der
oil f

Method:

Combine tapioca chips with all ingredients expect oil, and leave to marinate for about an hour.
Heat oil, deep fry chips till golden brown.

MURUKKU

Ingredients:

600 g gram flour *(besan)*
1 tbsp omum seeds – sieve and remove stones
1 slightly heaped tbsp chilli powder
2 slightly heaped tsp turmeric powder
1 tsp salt, mix with 300 ml water
oil for deep frying

Method:

Combine all the above dry ingredients in a large bowl, add salt water and mix till dough is smooth.
Put a handful of dough into murukku container.
Heat oil and lower murukku strands into hot oil by turning handle of the container. Fry till crisp.
Repeat till all the dough is used up.
Store in air tight container.

MINCED MEAT TART

Ingredients for filling:

120 g (3 tah) minced beef, lamb or pork
4 cloves garlic) grind to a fine paste
1 thick slice ginger 2 cm (¾in))
½ tsp black pepper
3 heaped tsp meat curry powder (see pg. 6)
salt to taste
6 sprigs big coriander leaves –chop
2 tbsp oil

Ingredients for pastry:

230 g (6 tah) plain flour
180 g (4½ tah) margarine
2 egg yolks
1-2 drops yellow food colouring – (optional)

Method for filling:

Combine minced meat with ginger-garlic paste, pepper, curry powder and salt.
Heat oil, add meat, cook till almost dry then add coriander leaves.
Stir and remove from heat. Leave to cool.

Method for pastry:

Sieve flour into a bowl. Rub margarine into flour, add egg yolks and colouring and bind to form a dough.
Take half the dough, roll out pastry to 0.5 cm (¼ in) thick on floured board.
Cut into rounds with a tart cutter. Fill centre with a teaspoonful of cooked meat. Repeat with the rest of the dough.
Bake in a moderately hot oven for 20 minutes.

CHICKEN KEBAB

Ingredients:

600 g (1 kt) chicken – separate meat from bones and cut meat into small cubes

Grind to a fine paste:

1 tsp cumin *(jintan puteh)* seeds) soak for ½ hour
1 tsp black peppercorns)
2 slices ginger – each 1 cm (½ in) thick
3 cloves garlic
2 red chillies

1 tsp coriander (*ketumbar*) powder
salt to taste
1 tbsp plain yoghurt or juice of ½ lemon
3 tbsp oil

Method:

Combine ground ingredients, coriander powder, salt, and yoghurt or lemon juice with chicken. Leave to marinate for about 10 minutes.
Heat oil, then cook till chicken is tender and fairly dry.
Alternatively, put marinated chicken on a tray brush with oil and grill.

To serve:

Put chicken pieces with pickled onions through cocktail sticks or little skewers.

SAVOURY FRIED TAPIOCA CHIPS

Ingredients:

600 g (1kt) fresh tapioca – peel and cut into chips
2 tbsp coriander (*ketumbar*) seeds) soak and grind
1 tsp cumin *(jintan puteh)* seeds)
1 tsp turmeric powder
salt to taste
oil for deep frying

Method:

Combine tapioca chips with all ingredients expect oil, and leave to marinate for about an hour.
Heat oil, deep fry chips till golden brown.

BANANA CRISPS

Ingredients:

2 bananas *(pisang kari)*

Combine to form a paste:

- 1 heaped tbsp chilli powder
- 1 level tbsp turmeric powder
- salt to taste
- 2 tbsp water

oil for deep frying

Method:

Peel bananas, cut them into very thin slices and soak in salted water for a few minutes. Drain, then mix the banana slices with the chilli-turmeric paste. Deep fry till lightly brown and crisp. Store in air-tight container.

Note:

Pisang kari are available at Changi, Geylang Serai or Kandang Kerbau Market.

VEGETABLE PUFFS
(Epok-Epok Sayor)

Ingredients for filling:

Grind to a fine paste:

- 1 clove garlic
- 2 red chillies
- 4 shallots

150 g (¼ kt) shrimps — remove shell, cut into small pieces

300 g (½ kt) bean sprouts — remove tails

10 sprigs coarse chives (*kuchai*) — chop fine

2 hard beancurd (*taukwa*) — deep fry till a light brown then cut into small cubes

salt to taste

2 tbsp oil

Ingredients for pastry:

340 g (10 tah) plain flour

5 tbsp cooking oil

¼cup water — mix with a little salt

Method for filling:

Heat oil, add ground ingredients and salt and fry till fragrant. Add shrimps, fry for a minute or two, then add bean sprouts and *kuchai.* Stir for a few minutes. Finally, add *taukwa* and stir, mixing vegetables well. Remove from heat, drain and cool.

Method for pastry:

Sieve flour into bowl, add oil and work it into the flour, adding just enough water to form a dough. Knead for about 5 minutes.

Roll out half the pastry at a time to 3 mm (1/8 in) thick. Use a 6 cm (2½ in) cutter to cut pastry into rounds.

Put a scant teaspoon of filling in each round. Damp edges of pastry with water, bring together, seal tightly and flute edges.

Deep fry till golden brown.

Serve with chilli sauce.

FRIED IKAN BILIS

Ingredients:

150 g (¼ kt) dried whitebait *(ikan bilis)*— remove head and slash each halfway down; do not wash

oil for deep frying

15 dried chillies — soak and grind to a fine paste

3 tsp sugar

salt to taste

2 tsp local vinegar

Method:

Heat oil, add *ikan bilis* and deep fry till crisp. Remove and drain.

Leave 8 tbsp oil in pan. Heat, add ground chillies and fry, then add sugar and salt and continue frying for a few minutes. Add vinegar and keep frying till mixture is fairly thick and well-cooked.

Add *ikan bilis,* stir, mixing well with chilli mixture. When thoroughly mixed, remove and put the *ikan bilis* in a strainer to drain off excess oil.

SPICY CUTTLEFISH

Ingredients:

600 g (1kt) small to medium sized cuttlefish *(sotong)* — clean well

2 tsp black peppercorns — grind fine

1½ tsp turmeric powder

salt to taste

3 tsp water

8 tbsp oil

Method:

Mix black pepper, turmeric powder and water and fry in oil until fragrant.

Add cuttlefish and salt and fry for about 5 minutes.

Remove and drain.

SATAY FLAVOURED COCKLES AND FISH BALLS

Ingredients:

600 g (1 kt) cockles *(kerang* or *see hum)* — scrub clean, scald and remove from shell
20 small fish balls
Grind to a paste: (neither too fine nor too coarse)
- 3 dried chillies — soak beforehand
- 1 stalk lemon grass *(serai)*
- 1 slice *lengkuas* — 1 cm (½in) thick
- 5 shallots
- 1 clove garlic
- 1 small piece dried turmeric — size of almond
- 1 tbsp coriander *(ketumbar)* seeds
- ½ tsp fennel *(jintan manis)* seeds
- ½ tsp cumin *(jintan puteh)* seeds

1 tsp tamarind) mix and strain to obtain juice
1 tsp water)
1 tbsp sugar
salt to taste
2 tbsp oil

Method:

Heat oil, then add ground ingredients and tamarind juice and fry till fragrant. Add sugar and salt and cook for a minute or two till sugar dissolves. Add fish balls and finally, the cockles.
Do not overcook as the cockles will become rubbery.

PRAWN FRITTERS

Ingredients:

600 g (1kt) medium sized prawns — remove shell but keep tails
3 slightly heaped tbsp rice flour
salt to taste
2½ tbsp water
oil for deep frying

Ingredients for gravy:

Grind to a fine paste:—
- 1 clove garlic
- 1 piece ginger — size of almond
- 15 dried chillies — soak beforehand

4 tbsp oil
3 tsp sugar
salt to taste
2 tsp light soya sauce
2 tsp vinegar

Method:

Put flour and salt in a bowl and mix with water to make a batter.
Add prawns to mixture.
Heat oil, add prawn and fry. Remove as soon as fritters are lightly brown and keep aside.

Method for gravy:

Heat oil, add ground ingredients and fry. Add sugar and salt, stir, then add soya sauce and vinegar. Continue stirring until gravy is fairly thick, then add fried prawn fritters and stir, mixing well with chilli mixture. Remove from heat as soon as fritters are well covered with gravy.

SAMOSAS

Ingredients for filling:

300 g (½kt) minced beef
½ cup water
salt to taste
1 tsp pepper
Grind to a fine paste:
- 5 cloves garlic
- 5 green chillies
- 2 red chillies
- ginger — 1 piece measuring 3x2x1 cm

3 onions — either grate fine or cut into small squares
1 tsp coriander *(ketumbar)* powder

Method for filling:

Combine minced beef, water, salt, pepper and chilli-garlic-ginger paste in a saucepan and boil till almost dry. Add chopped onions, coriander, chilli and turmeric powder, stir and remove from heat. Add chopped mint and coriander leaves, mix and leave to cool.

Method for pastry:

Sieve flour and baking powder into a bowl.
Melt the margarine and when cool, work into flour.
Add a little water at a time, using just enough to form a dough. Knead so that pastry becomes elastic.
Divide pastry into little lumps of equal size. Take 1

½ tsp chilli powder
½ tsp turmeric powder
4-5 sprigs big coriander leaves — chop fine
1 small bunch mint — chop fine

Ingredients for pastry:
340 g (10 tah) plain flour
a pinch of baking powder
120 g (3 tah) margarine
water

lump, roll it paper thin into a circle, dusting with flour to make rolling easier.
Cut each circle in half, place a spoonful of cooked meat mixture on each semi-circle; fold pastry over to form a triangle, moisten edges and seal carefully. Repeat till all the pastry is used.
Heat sufficient oil for deep frying and fry the samosas oven moderate heat till golden brown. Drain and serve with chilli sauce, mint sauce or yoghurt.

Note:
Size of samosas may vary; for cocktails, however, small ones are preferable.

Sweets and Cakes

BADAM HALWA
(Almond Sweet)

Ingredients:
230 g (6 tah) sugar
6 tbsp water
180 g (4½ tah) ground almonds
180 g (4½ tah) butter or *ghee*
2 tsp rose water
6 cardamons – remove seeds from pod and crush them
pinch of salt

Method:
Dissolve sugar in water over low heat, then bring to the boil.
Add almonds, rose water, butter or *ghee* and salt, then keep stirring.
When mixture thickens, pour into a shallow dish and sprinkle the crushed cardamom seeds on top.
Leave to cool before cutting into squares or diamond shapes.
Decorate with cherries.

SEVIAN OR SEMIAH

Ingredients:
6 tbsp *ghee*
120 g (3 tah) cashew nuts
120 g (3 tah) sultanas
150 g (¼ kt) yellow vermicelli – break into 3 pieces
1½ tins evaporated milk (tin size 410g/369 ml) – dilute with 3 cups hot water
230 g (6 tah) granulated sugar
1 tbsp rose water

Method:
Heat *ghee* in large, heavy saucepan. Deep fry cashew nuts then the sultanas separately.
Drain and keep aside.
Fry vermicelli in remaining *ghee* till brown, then add diuluted evaporated milk. Keep stirring, then add sugar and rose water. Bring to the boil then continue to boil for a few more minutes. Add sultanas and cashew nuts, cover, reduce heat and cook till vermicelli is soft. Serve hot.

1. *Badam Halwa*
2. *Gajjah Ki Halwa*
3. *Gulab Jamun*
4. *Sevian or Semiah*

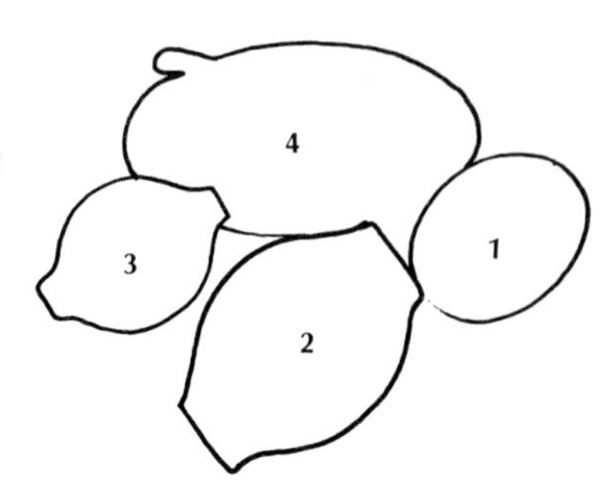

GAJJAH KI HALWA
(Carrot Sweet)

Ingredients:

600 g (1 kt) carrots – peel and grate coarsely
750 ml (3 cups) fresh milk
450 g (¾ kt) granulated sugar
60 g (1½ tah) almonds – scald, remove skin and chop
120 g (3 tah) cashew nuts – cut each in half and fry in 2 tbsp *ghee* till golden brown
2 tbsp rose water
3 tbsp *ghee*
120g (3 tah) sultanas
10 cardamoms – slit pods, remove seeds and discard pods

Method:

Put grated carrots and milk in a large, heavy saucepan and cook over high heat.
When liquid is reduced by half, add sugar, *ghee* and rose water, and continue cooking, stirring now and then.
When sugar has dissolved, lower heat, add almonds and cook for about 5 minutes. Add sultanas and continue to cook, stirring frequently.
When almost dry, add cashew nuts. When the mixture is dry and leaves the sides of the pan, remove from heat and sprinkle with cardamom seeds.
Spread mixture in a sandwich tin or cake tray and a allow to cool, then cut into squares or diamond shapes.
Alternatively, put a tablespoonful of mixture in the centre of a rectangular piece of cellophane paper, decorate with a cherry and almond slivers and wrap to form a neat package. Repeat till all the mixture is used.

Note:

The whole cooking process takes about an hour.

GULAB JAMUN
(Milk Sweets in syrup)

Ingredients:

4 cups milk powder
1 cup self-raising flour
4 tbsp *ghee*
250 ml (1 cup) fresh milk
ghee for deep frying

Ingredients for syrup:

3 cups sugar
3 cups water
1 stick cinnamon 3 cm (1¼ in) long
5 cloves
5 cardamoms

Method:

Combine all ingredients except *ghee* for frying in a bowl. Make little balls the size of walnuts.
Heat *ghee* then fry milk balls over low heat till dark brown. Drain, cool a little, then put milk balls into syrup.
Serve when cold.

Method:

Put all ingredients in a saucepan and boil till sugar dissolves.
Remove and cool a little before putting in the milk balls.

RICE DOUGH

If rice dough is not available at the market, it can be easily made with the use of a liquidiser or blender. To get the dough really fine and smooth may take a bit of time, causing the liquidiser to get heated. It is therefore suggested that after every 5 minutes the liquidiser should be switched off to allow it to cool for a couple of minutes.

Ingredients:

600 g (1 kt) broken rice (*beras hancor*) – clean, wash and soak 4–5 hours or preferably overnight
2 cups water

Method:

Put half the amount of rice and water in liquidiser. Liquidise till very fine; repeat with rest of rice and water.

PULOT SERIKAYA

Ingredients for first layer:
600 g (1 kt) glutinous rice *(beras pulot)*— wash and soak overnight in water to which a little salt has been added
1.35 kg (2¼kt) grated coconut – without skin. Extract 2 cups thick milk without adding water and reserve this for 2nd layer; then add 1 cup water to coconut and extract 1 cup milk for 1st layer
1 fragrant screwpine leaf *(daun pandan)*
salt

Ingredients for second layer:
300 g (½ kt) sugar
5 eggs
4 fragrant screwpine leaves *(daun pandan)* – cut into small piece and pound to extract juice
1 tsp green food colouring
¼ tsp vanillin crystals, or vanilla essence (optional)
3 level tbsp plain flour – sieve
2 cups thick coconut milk

Method:
Boil water in steamer.
Drain rice and put in steaming tray or sandwich tin, approx. 25 cm (10 in) diameter for a round tray or 20 cm (8 in) square for a square tray.
Add coconut milk, pinch of salt, *pandan* leaf broken in two. Stir then steam. After 20 minutes, turn rice over with a fork and continue steaming. After another 10 minutes, take tray out and press rice down with a piece of folded banana leaf till flat and compact. Steam for 5 more minutes.

Method:
Combine sugar and eggs in a fairly large bowl, and beat till sugar dissolves. Add flour, colouring, vanillin, juice from *pandan* leaf and mix well, then add coconut milk, stir and strain mixture.

To combine first and second layer:
When the first layer is cooked, prick the rice all over with a fork to ensure that the second layer adheres to it.
Pour the mixture for the second layer slowly over the first and steam for 10 minutes.
Remove lid, gently stir the top layer which is beginning to solidify, mixing it with the liquid, then continue steaming for 30 minutes.
When top layer is firm, remove and allow to cool well before cutting.

PUTU BAMBOO

Ingredients:
230 g (6 tah) rice flour
1 tsp salt
3 tbsp water
1½ cups grated coconut (remove skin) – set aside ½ cup

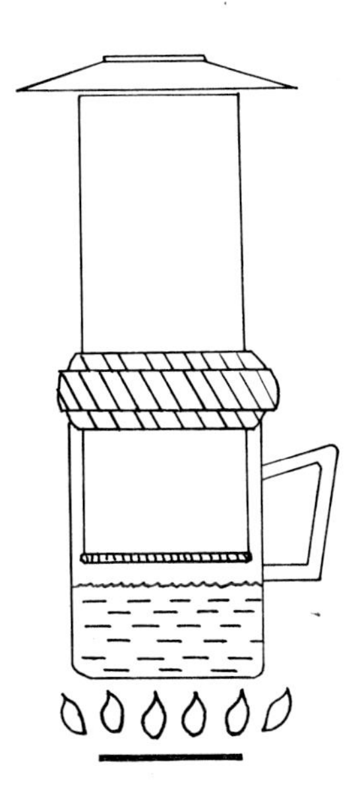

Method:
Dissolve salt in water and sprinkle over rice flour; mix well to get a crumbly mixture; add 1 cup grated coconut to rice mixture and mix well; put 1 tbsp grated coconut (from ½ cup set aside) into bamboo followed by 3 heaped tbsp of rice mixture; repeat till rice mixture is used up, making sure the last layer is of coconut; fit the bamboo snugly over a mug or kettle of boiling water, cover bamboo and steam until vapour is seen escaping from cake; when cooked, remove cake from bamboo by pushing it through, using the thick round handle of a wooden spoon or any similar object.
Serve with bananas and brown sugar.

KUEH KESWI

Ingredients:

300 g (½ kt) palm sugar *(gula Melaka)*
150 g (¼ kt) brown sugar
150 g (¼ kt) granulated sugar
400 ml water (for boiling sugar)
2 fragrant screwpine leaves *(daun pandan)*
600 g (1 kt) rice dough (see p. 102)
5 heaped tbsp tapioca flour
½ tsp lime *(kapor)* (see glossary) – mix with a little water into a paste
½ tsp salt
500 ml (2 cups) water
230 (6 tah) grated coconut (remove skin)) mix
1 tsp salt)

Method:

Boil water in steamer.
Put *gula Melaka,* brown sugar, granulated sugar, water and *pandan* leaves in a saucepan and heat until sugar dissolves, stirring occasionally.
Strain and cool.
Put rice dough, tapioca flour, *kapor,* salt and water in a large bowl and mix to a smooth consistency. Add to luke-warm syrup, stir well then strain. Fill either individual small moulds or deep trays with mixture. (Small Chinese teacups can be used as moulds; they should be heated in steamer for about 2-3 minutes before using). Steam *kueh* for *not more than* 20 minutes. When cooked, stand moulds in a shallow tray of cold water. When *kueh* are cold, remove from moulds by loosening edge with a knife.
Serve with grated coconut which has been mixed with salt.
Makes about 75.

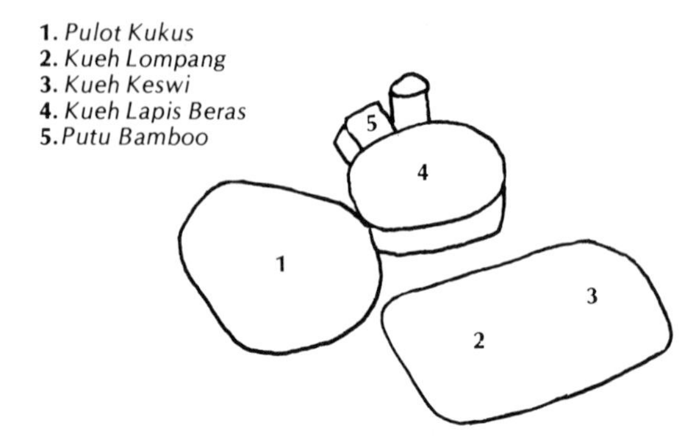

1. *Pulot Kukus*
2. *Kueh Lompang*
3. *Kueh Keswi*
4. *Kueh Lapis Beras*
5. *Putu Bamboo*

KUEH LAPIS BERAS

Ingredients:

450 g (¾ kt) sugar
2 cups water (for boiling sugar)
2 fragrant screwpine leaves (*daun pandan*)
900 g (1½ kt) grated coconut (remove skin) – for *first squeeze,* add 1 cup water; using a fine tea towel or piece of muslin, extract 2 cups thick milk. For *second squeeze,* add 1 cup water to coconut and extract equal amount of milk
600 g (1 kt) rice dough (see p. 102)
4 heaped tbsp tapioca flour
pinch of salt
½ tsp cochineal

Points to note:

When making *nonya kueh* such as *Kueh Lapis Beras,* to obtain thick, rich milk from coconut, use either a fine tea towel or a piece of muslin to squeeze coconut.
Use a fine mesh strainer to strain mixture before steaming. If rice *flour* is used instead of rice *dough,* the *kueh* will not be fine in texture.
Make sure that each layer is firm before adding the next. When one layer is opaque or milky looking, not transparent, then add the next layer, pouring the mixture in *gently.*

Method:

Boil water in steamer.
Put sugar, water and *pandan* leaves in a saucepan and heat till sugar dissolves.
Leave syrup until it is luke warm, then add to coconut milk (first squeeze). Combine rice dough with tapioca flour and salt. Add coconut milk (second squeeze) and mix to a smooth paste, then add coconut milk-syrup. Stir till smooth and well mixed then strain with fine mesh strainer. Grease a square or round (27 cm or 11 in diameter) sandwich tin. Heat tin in steamer for about 5 minutes.
Put 4 cups of mixture in a bowl and colour with cochineal. Leave the rest of the mixture white.
Pour 1 cup of white mixture into the heated tray and steam for 6–8 minutes, then add ¾ cup red mixture over first layer, pouring it gently. Steam for 5 minutes.
Repeat process, steaming white and red layers alternately till all the mixture is used up. For the last layer, which should be red, pour 1 cup mixture and steam for 10–12 minutes.
Cut when thoroughly cool.

PULOT KUKUS

Ingredients:

For the filling:
- ¾ cup brown sugar
- ½ cup water
- 1 cup grated coconut

2 cups glutinous rice flour *(tepong pulot)*
230 g (6 tah) grated coconut – add ½ cup water and, using a fine tea towel squeeze coconut to extract ½ cup thick coconut milk
½ tsp salt
½ tsp green food colouring
banana leaf cut into 20 rectangular pieces (4 x 5 cm or 1½ x 2 in)

Method:

For the filling:
Put sugar and water in a saucepan and heat till sugar dissolves. Strain and put back in saucepan. Add grated coconut and cook over low heat, stirring all the time, till mixture is of a thick and moist consistency. Cool.

Add colouring to coconut milk. Combine glutinous rice flour, salt and coconut milk and mix into a soft dough. Take a small lump of dough, flatten it a little, put in coconut filling and wrap dough around it to form an oval shape about 4-5 cm (1½ - 2 in) long. Make design with a tart pincher or trace design with a skewer. Put each on a piece of banana leaf and steam for about ½ hour.
Serve while still warm. Makes about 20.

KUEH KOCHI KUAH

Ingredients for the filling:

1 cup grated coconut – with or without skin
1 cup brown sugar
¼ cup water

Ingredients for glutinous rice dough:

1½ cups glutinous rice flour *(tepong pulot)*
¼ tsp green food colouring
1 tsp salt

Method for filling:

Combine ingredients and cook over low heat till thick, stirring all the time. Set aside to cool then shape into little balls, the size of small marbles.

Method for rice dough:

Put the glutinous rice flour in a bowl. Add colouring, juice of *pandan* leaves, salt and enough water from the half cup allowance to form a dough. Take a small

½ cup water
2 fragrant screwpine leaves – pound and extract juice

Ingredients for sauce:
230 g (6 tah) grated coconut – without skin
add 2 cups water and extract 2 cups coconut milk
2 level tbsp plain or rice flour
2 fragrant screwpine leaves *(daun pandan)*

lump of dough, flatten, put a ball of filling in the middle and wrap dough round it, shaping it into a ball. In a saucepan, heat water to which a little salt and one *pandan* leaf has been added. When water is boiling, drop in the balls of dough. When they rise to the surface, they are cooked. Remove, drain and put them in the sauce.

Method for sauce:
Combine the above ingredients and bring to the boil, stirring all the time. When fairly thick, remove from heat. Cool a little.

Ingredients for sugar coating:
60 g (1½ tah) granulated white sugar or brown sugar
2 tbsp water

Method:
Cook water and sugar over high heat. When thick and syrupy, put the deep fried cakes into it and keep turning till cakes are covered with syrup.
Keep stirring till sugar crystallizes, then remove from heat.

COCONUT CRUMBS

These cooked crumbs may be used for garnishing Koleh Koleh as well as for Mee Siam Lemak.

Ingredients:
450 g (¾kt) grated coconut – add 2 cups water and extract equal amount of milk

Method:
Put the coconut milk into a pan and boil over low heat. When liquid has been absorbed and all that remains is oil and crumbly coconut, keep stirring. When the coconut crumbs are brown, remove from pan.
The oil can be kept aside for brushing over banana leaves for Kueh Kochi Daun.
The crumbs can be stored in an air tight bottle.

KUEH JONKONG

Ingredients:
2 cups rice flour
2 level tbsp tapioca flour
680 g (1 kt 2 tah) grated coconut – add 4 cups water and extract 6 cups milk
1 block palm sugar *(gula Melaka)* – cut into little slivers
1 fragrant screwpine leaf *(daun pandan)*
1 tsp salt

Method:
Combine rice flour, tapioca flour, coconut milk and salt. Mix and strain into a saucepan.
Break *pandan* leaf into mixture then cook over low heat, stirring all the time. When the mixture becomes fairly thick, remove from heat.

To wrap into bundles:
Have ready about 12 to 15 pieces of banana leaves each measuring 25cm x 15cm (10 x 6 in)
Put two tablespoons of cooked mixture on banana leaf. Put a heaped teaspoon of **gula Melaka** in the centre of the mixture, put another tablespoon of mixture over this and fold bundle in this way:

bring the two lengths of the leaf towards the centre so that they overlap, covering the mixture ; tuck the two ends underneath.

Put bundles in steamer (boil water in steamer beforehand) and steam for 20 minutes. Serve cold.

KUEH LOMPANG

Ingredients:

450 g (¾ kt) granulated sugar
500 ml (2 cups) water (for boiling sugar)
2 fragrant screwpine leaves *(daun pandan)*
600 g (1 kt) rice dough (see pg. 102)
5 level tbsp tapioca flour
500 ml (2 cups) water
½ tsp salt
food colouring — 3-4 colours
230 g (6 tah) grated coconut (remove skin))
1 tsp salt) mix

Method:

Boil water in steamer.
Put *pandan* leaves, sugar and water in a pot and heat until sugar dissolves. Strain and cool.
Put rice dough, tapioca flour, salt and water in a large bowl, and mix to a smooth consistency.
Pour luke-warm syrup into rice mixture and stir well, then strain. Put mixture in 3 or 4 bowls of the same size (depending on the number of colours you plan to use) add a drop or two of colouring to each bowl of mixture and stir well. Pour mixture into little Chinese teacups (one size smaller than those used for *kueh keswi*) which have been pre-heated in steamer and steam for *not more than 20 minutes.*
Repeat steaming process till all the mixture has been used, making sure each time that the teacups are heated in steamer before mixture is poured into them. When cooked, stand moulds in a shallow tray of cold water; when *kueh* are cold, remove from moulds by loosening edge with a knife.
Serve with grated coconut wich has been mixed with salt.
Makes about 90-100.

1. *Kueh Kochi Kuah*
2. *Pulot Serikaya*
3. *Rice Pudding with White Syrup*
4. *Getas Getas*
5. *Kueh Jonkong*

RICE PUDDING WITH WHITE SYRUP

Ingredients:
1 cup rice flour
450 g (¾ kt) grated coconut – add 2 cups water and extract 4 cups milk
1 level tbsp tapioca flour
1 tsp salt
2-3 fragrant screwpine leaves *(daun pandan)*

Method:
Combine rice flour, coconut milk, tapioca flour and salt.
Stir and strain into a saucepan.
Add *pandan* leaves and cook, stirring all the time.
When thick, remove from heat, put into individual moulds or into a tray and chill.
When serving, pour syrup over pudding.

Syrup Ingredients:
1 cup granulated sugar
2 cups water
1 fragrant screwpine leaf *(daun pandan)*

Method:
Combine the above ingredients in a saucepan and bring to the boil.
Cool before pouring over pudding.

GETAS GETAS

Ingredients:
2 cups glutinous flour *(tepong pulot)*
1 cup grated coconut – without skin
1 tsp salt
75 ml water
oil for deep frying

Method:
Combine rice flour, coconut, salt and water in a bowl and mix to form a dough. Shape into ovals about 5 cm (2 in) in length.
Deep fry over low heat till golden brown.

Ingredients for sugar coating:
60 g (1½ tah) granulated white sugar or brown sugar
2 tbsp water

Method:
Cook water and sugar over high heat. When thick and syrup, put the deep fried cakes into it and keep turning till cakes are covered with syrup.
Keep stirring till sugar crystallizes, then remove from heat.

1. *Koleh Koleh*
2. *Tepong Gomak*
3. *Kueh Kochi Lemper*
4. *Chicken Lemper (back of plate)*

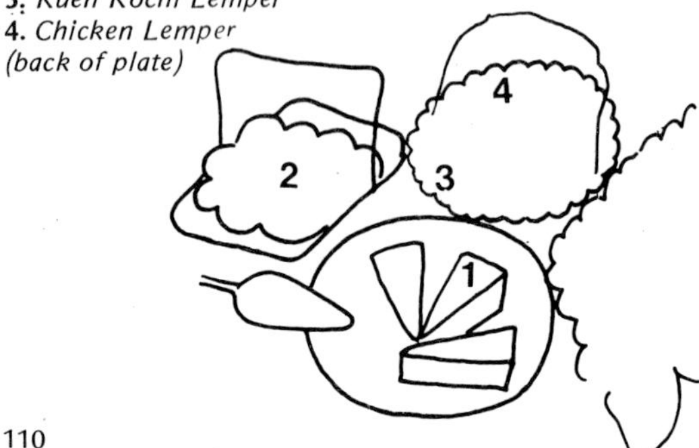

KUEH KOCHI DAUN

To wrap the *Kueh,* you will need 30 pieces of banana leaves each measuring about 8 cm x 10 cm (3 in x 4in) – wash and wipe them.

Ingredients for filling:
1 cup grated coconut
½ cup brown sugar
¼ cup water
1 fragrant screwpine leaf *(daun pandan)* – tie into a bundle
pinch of salt

Method:
Combine all the above ingredients in a pot and cook over low heat, stirring all the time, till most of the moisture is absorbed and the mixture has a moist consistency. Leave to cool.

Ingredients for Kueh:
2 cups glutinous rice flour *(tepong pulot)*
230 g (6 tah) grated coconut (without skin) – add ¾ cup water and extract equal amount of milk
½ tsp salt

Method:
Combine all the above ingredients in a bowl and mix to form a dough.

To wrap:
Take a piece of banana leaf, brush some oil over it, place a small lump of dough in the centre and flatten. Put a teaspoonful of coconut filling in the centre and wrap dough around filling. Bring the two lengths of the leaf towards the centre so that one overlaps the other, and tuck the two ends under. Repeat till all the dough is used up.

To cook:
Steam for 20 minutes
This will make about 25–27 *Kueh.*

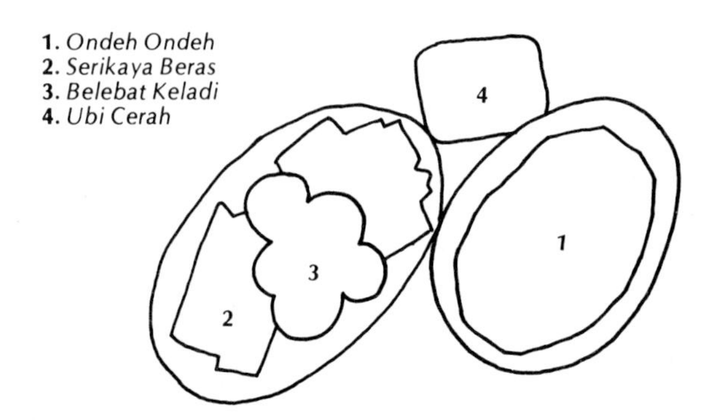

1. *Ondeh Ondeh*
2. *Serikaya Beras*
3. *Belebat Keladi*
4. *Ubi Cerah*

BELEBAT KELADI
(Steamed Yam Cake)

Ingredients:

450 g (¾ kt) yam – peel then grate
1 cup grated coconut – without skin
1 heaped tbsp rice flour
6 heaped tbsp granulated sugar
pinch of salt
drop of violet food colouring (optional)
banana leaves – cut into 10 rectangular pieces measuring 30 x 15 cm (12 x 6 in)

Method:

Combine rice flour with yam in a bowl. Add coconut, sugar, salt and colouring, and mix well.
Put a heaped tablespoon of yam mixture in the centre of the banana leaf. Bring the two long ends to the centre so that they overlap and tuck the other two ends under. Repeat till all the mixture is used.
Steam for 30 minutes. (Makes about 8–10 packets)

SERIKAYA BERAS
(Steamed Rice-Custard Cake)

Ingredients:

5 eggs
230 g (6 tah) sugar
450 g (¾ kt) grated coconut – without adding water, squeeze coconut to obtain 1 cup thick milk
2 slightly heaped tbsp rice flour
3 - 4 fragrant screwpine leaves *(daun pandan)*
pinch of salt
2 drops green food colouring
pinch of vanillin crystals, or drop of vanilla essence (optional)

Method:

Boil water in steamer.
Beat eggs and sugar in a bowl till sugar dissolves, then add coconut milk and stir. Add rice flour, stir, then add juice of pandan leaves, salt, colouring and vanilla. Mix well and strain mixture into a sandwich tin. Steam for 10 minutes then stir. Cover and steam for another 20 minutes.
Allow to cool before cutting.

ONDEH ONDEH KELEDEK
(Sweet Potato Balls)

600 g (1 kt) sweet potatoes – boil in water to which 2 fragrant screwpine leaves *(daun pandan)* have been added, peel then mash while still warm
4 fragrant screwpine leaves *(daun pandan)* – pound to extract juice
4 heaped tbsp plain flour
1 block palm sugar *(gula Melaka)* – cut into little slivers
230 g (6 tah) grated coconut – (remove skin) steam with a little salt to prevent the coconut from going sour

Method:

Squeeze juice from crushed *pandan* leaves into mashed sweet potatoes. Add flour and a pinch of salt and mix.
Take a small lump of dough, flatten it, put a few slivers of *gula Melaka* in the middle and roll into a ball. Repeat till all the dough is used up.
Boil a pan of water. Drop sweet potato balls into boiling water.
Remove when they rise to the surface, drain and roll in steamed coconut.

TEPONG GOMAK

Ingredients for filling:

½ cup dried green beans – soak for 2 hours
2 cups water
1½ cups granulated sugar
½ cup grated coconut

Ingredients for dough:

2 cups glutinous rice flour *(tepong pulot)*
½ tsp salt
¾ cup water

Method:

Boil the beans in the water till very soft. (A pressure cooker would hasten the cooking.)
When beans are soft, add granulated sugar and grated coconut, and cook over low heat stirring at first, to mix ingredients, then leave to cook till all the moisture is absorbed and the mixture has the consistency of a puree. Leave to cool.

Method for dough:

Combine the above ingredients and mix into a dough. Take a lump of dough (size of a walnut), flatten, put in a teaspoonful of green bean filling. Wrap dough around it and shape into a flattish cake.

Note:
For the filling, the bigger green beans are preferable to the small ones.
See recipe for Koleh Koleh for the preparation of green bean flour.

To cook:
Boil a pan of water to which a pinch of salt and 1 *pandan* leaf have been added. When water is actively boiling, drop in the cakes. When they rise to the surface, they are cooked.
Roll them over some green bean flour till they are evenly coated.
(Makes about 2 dozen)

UBI CERAH
(Tapioca Chips Coated in Gula Melaka)

Ingredients:
600 g (1 kt) fresh tapioca – peel, wash and slice very thin
1 block palm sugar *(gula Melaka)* – cut into pieces
5 tbsp water
oil for deep frying

Method:
Rub tapioca slices with salt then deep fry till light brown; drain and cool.
Put *gula Melaka* and water in a pan and dissolve over low heat, then add the tapioca slices and stir till they are well coated with syrup. Remove from heat and leave to cool before storing in an air tight container.

KOLEH KOLEH

Prepare the *green bean flour* in advance. For this, you will need 300 g (½ kt) of dried green beans. Remove dust, stones and any other impurities, then roast over heat till slightly brown. Grind into flour.

Ingredients for Syrup:
230 g (6 tah) palm sugar *(gula Melaka)*
½ cup water
6 tbsp granulated sugar
1 fragrant screwpine leaf (*daun pandan*)

Ingredients for Koleh Koleh:
1 cup dried green bean flour
450 g (¾ kt) grated coconut – add 2½ cups water and extract equal amount of milk
pinch of salt

Method for syrup:
Combine all the ingredients in a pot. Put over heat to dissolve the sugar, then leave to cool.

Method for Koleh Koleh:
Combine flour, prepared syrup, coconut milk and salt in a large bowl. Mix well, then strain into a tray and steam. Five minutes after putting it in, remove cover, stir mixture then cover and continue steaming for another five minutes. Remove cover and stir again, then cover and continue steaming for 30–35 minutes.

Alternative Method:
After steaming for 5 minutes, stir for 20 minutes over moderate heat till mixture thickens, then cover and leave to steam for another 10-15 minutes.
Sprinkle coconut crumbs over Koleh Koleh before serving.

LIST OF SUPPLIERS

MONTREAL.

World Wide Imported Foods., 6700 Cote Des Neiges, Montreal, Quebec, Canada.
514 – 733 – 1463.

S. Enkins Inc., Imports and Exports,
1201, St. Lawrence, Montreal 18, Canada.

VANCOUVER

Famous Foods Limited, 3839 Commercial Drive
Vancouver, B.C. (604) 872-3019

Patel's Supermarket, 2210 Commercial Drive
(at East 6th Avenue) Vancouver, B.C.
(604) 255-6729
"also at Overwaitea Stores in B.C."

TORONTO

Kensington Market
Situated between College and Dundas Streets, west of Spadina Avenue, this open-air, cosmopolitan market is a shopper's delight. The enjoyment of finding every imaginable delicacy is heightened by the European bazaar atmosphere unique to this area. Storekeepers are extremely helpful – particularly if you are unfamiliar with many of the products. The following stores are a few of the many found in this neighbourhood.

HOUSE OF SPICE
190 Augusta Avenue (416-)363-8544

ALVES WEST INDIES
70 Kensington Avenue (416-)368-3356

ASIAN FOOD COMPANY
187 Baldwin Avenue (416-)368-8841

LUSITANIA GROCERY
275 Augusta Avenue (416-)366-2123

SANCI TROPICAL FOODS CO.
66 Kensington Avenue (416-)368-6541

Chinatown
Just a few blocks south of Kensington Market, on Dundas Street is Toronto's Chinatown. The stores offer a wide variety of rice, noodles, beans, fruits, vegetables, and condiments. The names of ingredients may vary, but shopkeepers are helpful and will provide information and explanations.

St. Lawrence Market
Located at Front and Jarvis, the St. Lawrence Market offers the best variety of fish, cheese, fresh produce, and baked goods in the city.

Toronto – general
Throughout Toronto are many other fine stores in which you may find many of the ingredients for these recipes.

IWAKI JAPANESE FOOD STORE
2627 Yonge Street (north of Eglinton)
(416-)481-8928

SANKO TRADING COMPANY
221 Spadina Avenue
(416-)862-1082

JOHN WILSON IMPORTED FOODS
3249 Yonge Street (north of Lawrence Ave.)
(416-)488-2030

BOMBAY BAZAAR
4 Rexdale Boulevard (west of Islington)
(416-)742-2782

AFRO–ASIAN FOOD DISTRIBUTORS
918 Pape Avenue
(416-)463-1287

EAST/WEST INDIAN FOOD
1640 Jane Street
(416-)242-2505

INDO–CANADA GENERAL FOOD
624 Bloor Street West
(416-)536-2666

Toronto-general

INDIA PRODUCTS & GIFTS
1454 Queen Street West
(416-)533-4444

PHILIPPINE ASIAN TRADING COMPANY
482 Parliament Street (416-)967-1700

PHILHOUSE CANADA
495 Queen Street West (416)869-0917

PHILIPPINE ORIENTAL FOOD MARKET
1033 Gerrard Street East
(416-)924-5991

PILIPINO FIVE–O STORES
50 Gladstone Avenue.
(416-)535-2322
1427 Queen Street West
(416-)535-3676

T. EATON CO.
2010 Yonge Street.

Glossary

AGAR AGAR: a type of gelatine made from seaweed that sets without refrigeration. Available in strands, which must be soaked in water for ½ hour before use, or powder. Strands are used in this book, but powdered agar agar can be substituted (1 teaspoon will set approximately 2 cups water). If using powder, pre-soaking is unnecessary. Ordinary gelatine cannot be used as a substitute.

ASAFOETIDA: a gum derived from a Persian plant that is used in tiny amounts in Indian cooking to give flavour and to prevent flatulence. Available in small rectangular blocks from Indian spice shops; known as *hing* in Hindi and *perankayam* in Tamil.

ASAM: see Tamarind

BANANAS: In savoury dishes, non-sweet bananas similar to plantains are used (such as *pisang kepok, pisang kari, pisang nipah).* Be sure not to substitute with sweet dessert bananas. Banana leaf is frequently used to wrap cakes and fish. Foil can be substituted in most cases, although the flavour will be different.

BASMATI: fragrant long-grain rice usually from Pakistan, it has a delicious nutty flavour. Sometimes sold as *Patna* rice. Available in Indian shops and in some health food shops.

BEAN CURD: the soya bean is one of the most versatile products in Asia. Bean curd made from soya beans is available in several consistencies. The firm bean curd is known as *taukwa,* and is available in Chinese delicatessens overseas. Do not confuse with soft beancurd *(tauhu).* Dried bean curd skin, available in sticks *(foo chok)* or sheets *(tim chok)* is sold loose or in packages.

BEAN THREAD VERMICELLI: called *sohoon,* or sometimes *tunghoon,* these fine transparent noodles are also known as *fun* see. They are made from flour of the dried green*mung* bean.

BESAN: flour made from chick peas *(kacang kuda).* Sometimes called 'gram flour'. Do not confuse with flour made from yellow lentils.

BLIMBING: a small oblong acidic fruit used by Malay and some Chinese cooks. Green tomatoes could be used as a substitute.

BRIANI SPICES: ready-mixed spices for use in Nasi Briani sold by some Indian spice merchants, consisting of cinnamon, cloves, cardamom, and a spice known as 'black cumin' which is thinner than regular cumin and has a different flavour.

BROWN MUSTARD SEEDS: small seeds used by South Indian cooks, known as *biji sawi.* Do not confuse with the larger yellow mustard seeds from Europe — the flavour is quite different.

CANDLENUTS: a cream-coloured waxy nut *(buah keras)* used for texture in Malay and Indonesian dishes. Substitute macadamia nuts or cashews.

CHICK PEA FLOUR: see Besan

CHILLIES: unless otherwise specified in recipe, use fresh chillies of the variety that is about 10-15cm (4-6in) long. Birds-eye chilli *(chilli padi)* are about 2cm (¾in) long and *very* hot. *Chilli-tairu* are green chillies preserved in salted yoghurt; store in tins and deep fry in hot oil for a few seconds.

COARSE CHIVES: Flat dark green chives known as *kuchai,* available in Chinese delicatessens. Slightly different flavour to fine round chives; the green tops of spring onions would be an acceptable substitute.

COCONUTS: Freshly grated coconut is used to make coconut milk by adding water and squeezing to obtain the liquid. If you cannot obtain fresh coconuts, the following methods can be used to obtain coconut milk:

Creamed Coconut: a solid preparation sold in round plastic tubs. To obtain thick coconut milk, mix 100g (3½ oz) creamed coconut with 1 cup boiling water. Stir until dissolved and strain while still hot. For thin coconut milk, use 30g (1 oz) creamed coconut to 1 cup water.

Desiccated Coconut: blend 2 cups desiccated coconut with 2 cups water in a blender. Mix at high speed for 30 seconds, then squeeze and strain for thick coconut milk. Return coconut to blender and mix with another 2½ cups hot water and repeat the process to obtain thin coconut milk.
Cakes can be made using desiccated coconut moistened with a little milk if fresh coconuts are not available, although naturally the flavour is not quite as good.

CORIANDER LEAVES: known as *wan swee* to Chinese cooks, these leaves have a distinctive flavour. The 'small' coriander leaves specified in some recipes are actually the first leaves of the young coriander seedling, and are milder in flavour than the mature leaves. The Malay name for 'small' coriander is *daun ketumbar halus;* if the leaves are not available, use the 'big' leaves, *daun ketumbar kasar.* If you have difficulty obtaining fresh coriander leaves, try growing them in a pot from the coriander spice seeds that you buy in a spice shop.

CURRY LEAVES: used abundantly by South Indian cooks, curry leaves *(karuvapillai* in Tamil, *daun kari* in Malay) must not be confused with the Indonesian *daun salam,* a type of bay leaf. Curry leaves are about 2.5cm (1 in) long, dark green in colour, and have a pungent smell. Dried leaves can be bought overseas in specialist curry shops.

DAUN KESOM: a pungent dark green herb (bot: polygonum); nearest substitute is mint.

DRIED GREEN BEAN: this is the *mung* bean, a tiny green pea-shaped bean known as *lok tau* by Chinese cooks, and *kacang hijau* by Malays. Flour made from this bean is known as *tepong hoen kwe.*

DRIED LILY BUDS: sometimes called 'golden needles,' a direct translation of the Chinese name *kim chiam,* these are slender golden-brown strips about 8cm (3in) long.

DRIED SHRIMP PASTE: called *blacan* by Malays, *trassi* by Indonesians, this is a very strong-smelling seasoning which gives an authentic touch to many dishes.

DRIED WHITEBAIT: these range in size from the very tiny thin variety often known as 'silver fish' up to the large variety about 5cm (2in) long. If using large ones, discard head. Known as *ikan bilis,* these fish have been salted before being dried, so always taste a recipe using these before adding salt.

EGGPLANT: Malay name is *terong,* the Indian name is *brinjal.* Asian varieties are much smaller than European eggplant or aubergine, and are generally around 15—18cm (6-7in) in length.

FISH CURRY SPICES: also known as *rempah tumis ikan,* this is a mixture of whole spices including brown mustard seed, fenugreek, cumin, fennel and husked blackgram *dhal (biji sawi, alba, jintan puteh, jintan manis* and *urad dhal).* If you cannot obtain this ready-mixed from your spice shop, mix your own using 1 teaspoon of each spice except for fenugreek, which should be just ½ teaspoon.

FRAGRANT LIME LEAVES: a wonderfully fragrant leaf from a variety of citrus, known as *daun limau perut.* Any young citrus leaves could be substituted, although their flavour is nowhere near as lovely.

FRAGRANT SCREWPINE LEAF: a variety of *pandanus* palm known as *daun pandan* and used to flavour cakes and desserts. It is a long (at least 30cm or 12in), narrow, shiny leaf, easily grown in a tropical garden.

FRESH RICE FLOUR NOODLES: sometimes sold in a wide flat sheet which you must cut into strips with scissors, otherwise sold ready-cut. Known as *kway teow* or *sa hor fun.* Available dried from Chinese stores; if using the dried noodles, soak in boiling water for ½ hour, drain, and soak in a second lot of boiling water for another ½ hour.

GHEE: butter oil favoured by Indian cooks since it does not burn as readily as butter. Obtainable in tins from Indian stores. To make your own *ghee,* heat a packet of butter over very low heat until it melts. Allow sediment to sink to the bottom of the pan, then pour off the golden oil sitting on the top. Put oil into a clean saucepan, heat again, then strain through muslin to remove any trace of sediment. Keeps several months without refrigeration.

GLUTINOUS RICE: a variety of rice which becomes very sticky when cooked; used mostly for cakes and known as *pulot.* Generally available in Chinese stores.

KAPOR: white lime used as part of a *betel* quid. Obtainable in Indian food shops.

LEMON GRASS: a type of grass with a pungent lemony fragrance, this is used to make citronella. Known as *serai,* it grows like a small leek bulb. Use only the bottom 10-15 cm (4-6 in), discarding the tough upper leaves. Strips of dried lemon grass are sometimes sold under the Indonesian name *sereh.* Powdered *serai* or *sereh* can also be substituted; use about 1 teaspoon in place of 1 stalk of fresh lemon grass.

LENGKUAS: a ginger-like root of the galangal family. Powdered *lengkuas or laos* (the Indonesian name) is often obtainable in curry shops overseas; substitute 1 teaspoon of powder for 1 thick slice of fresh *lengkuas.*

LIMES: two types of limes can be used; the large green lime, which is shaped like a lemon, and the small round lime, which is known as *limau kesturi.* The small lime has more fragrance than the large variety; substitute half-ripe kumquats if possible, otherwise use lemon juice.

LOCAL VINEGAR: very strong synthetic vinegar used by many Singapore cooks. If using regular white vinegar such as Heinz, use only half the amount specified for local vinegar.

OMUM: a minute spice resembling parsley seed, this is used to flavour Murukku and some other Indian dishes. The botanical name is *carom,* the Hindi *ajwain.* No substitute.

PALM SUGAR: known as *gula Melaka,* this is a hard brown sugar made from the sap of the *aren* palm. If not available, substitute soft brown sugar with a touch of maple syrup.

PAPADUM: also spelled *popadum,* these are paper-thin wafers made from lentil flour. They should be absolutely dry before frying in hot oil for a few seconds on either side, so that they swell up and become crisp and golden.

RICE VERMICELLI: popularly called *beehoon,* this noodle is a very thin white thread made from rice flour. Sometimes called *mi fun* or rice stick noodles.

ROSE WATER: a wonderfully evocative flavour, this is Arabian in origin. Used in some Indian and Malay dishes. If using the concentrated rose essence, be sure to use much less than the amount required for rose water, which is diluted.

SALTED SOYA BEANS: soft brown salty beans in a thick paste, called *taucheo.* Sometimes sold in jars labelled 'bean sauce' in Western countries.

SARDINES: small mackerels, usually packed in tomato sauce, are misleadingly labelled 'sardines' in Malaysia and Singapore. They are much longer than, and different in flavour from European sardines, so if not using a Malaysian or Singaporean brand, buy mackerel, snoek or even herring in tomato sauce.

SHALLOTS: the Malay name, *bawang merah,* means 'red onion' and is a good description of these pinkish-purple, marble-sized onions. If shallots are not available, use big purple Bombay onions or, failing these, brown skinned onions. One big onion is roughly equivalent to 8-10 shallots.

SOYA SAUCE: two types are used: thin light soya sauce, and thick dark black soya sauce. As the flavours differ, be sure to use the one called for in the recipe.

SPRING ONIONS: slender green stalks with a white base. Do not confuse with shallots, which are small round pink onions.

TAMARIND: the dried pulp of the fruit from the tamarind tree *(asam)* is used to give fragrant sourness to many dishes. Sometimes slices of dried fruit *(asam keping or asam gelugor)* are used. Generally available in curry shops.

TURMERIC: the fresh or dried root of the turmeric plant is used for colour and flavour in many dishes. Dried powdered turmeric can be substituted: use about 1 teaspoon powder to 1 cm (½ in) fresh turmeric. Turmeric leaves are occasionally used as a herb; if these are not available, omit, as there is no substitute.

WHITE POPPY SEEDS: sand-like grains that are creamy white in colour, poppy seeds are used mainly as a thickening agent. If white poppy seeds *(kas kas)* are not available, try ground almonds.

YAM BEAN: a crunchy white vegetable called *bangkwang* with a flavour that is like a cross between an apple and a potato. Tinned water chestnuts are probably the best substitute.

Index

A

B

C

D

E

F

G

H

I

J

K

L

M

N

O

P

R

S

T

U

V

W

Y